T5-AXN-039

Quality Management Systems

A PRACTICAL GUIDE

Quality Management Systems

A PRACTICAL GUIDE

HOWARD S. GITLOW, Ph.D.

S^t_L

St. Lucie Press
Boca Raton • London
New York • Washington, D.C.

Library of Congress Cataloging-in-Publication Data

Gitlow, Howard S.   Cae
    Quality management systems : a practical guide / Howard S. Gitlow.
        p.   cm.
    Includes bibliographical references and index.
    ISBN 1-574-44261-9 (alk. paper)
        1. Total quality management. 2. Small business—Management. 3. Quality
control—Standards. 4. Organizational effectiveness. I. Title.

    HD62.15 .G538 1999
    658.4'013 21--dc21   99-043472
                                                                                 CIP

Visit the CRC Press Web site at www.crcpress.com

© 2001 by CRC Press LLC
St. Lucie Press is an imprint of CRC Press LLC

No claim to original U.S. Government works
International Standard Book Number 1-574-44261-9
Library of Congress Card Number 99-043472
Printed in the United States of America   2 3 4 5 6 7 8 9 0
Printed on acid-free paper

PREFACE

Do you remember the first time you drove a car? You probably do. You probably also remember the prelude to that drive: reading the driver's manual, watching movies, practicing in your driveway, and endless discussions about the impending event with your equally inexperienced friends. By the time you actually drove, you knew a lot about the "theory of driving." You just didn't know exactly how to translate that theory into practice. I assume you made that jump over the chasm between theory and practice, even if it was purely by force of bravado and terror.

Quality Management has become an integral part of business. It's time to put theory into practice. Since the early 1980s, interested people have read books and journals, attended seminars and training sessions, and watched films and videos in an earnest attempt to learn about Quality Management. Their education has been frustrating because although they embrace the theory, it is difficult for them to apply it in their organizations. What is needed is an integrated approach that explains the theory and how to put it into practice using a step-by-step method.

Quality Management Systems: A Practical Guide presents a model of Quality Management that combines the theoretical base of Dr. W. Edwards Deming and the practical techniques of the Japanese into a useful application for those people who have been wondering, "How do we get started and actually 'do' quality?" It is a fork-shaped model that includes the handle of the fork (Management's Commitment to Transformation), the neck (Management's Education) and three prongs (Daily Management, Cross-Functional Management, and Policy Management). To facilitate understanding, numerous organizational, personal, and familial examples are provided.

Management's Commitment to Transformation, the handle of the fork, is clearly the most essential element of any Quality Management undertaking. Management has to be ready to digest the theory of Quality

Management and put it into practice. This has not been an easy meal to swallow, and they are certainly not the only group that is reluctant to change. For example, surgeons were very hesitant to go from theory to practice. It was postulated that the washing of hands before surgery could prevent infection; however, it took surgeons 200 years before they put theory into practice and scrubbed up before operating. It is my sincere hope that managers will be more open-minded and enthusiastic about transformation.

Dr. Deming passed away on December 20, 1993. His dream was "to save America from committing suicide." His philosophy of management will endure. He never wavered in his mission to transform American management. For those of us who were lucky enough to have been touched by Dr. Deming, some goals remain: disseminating his theories, building on them, and working with people in all types of organizations to make "joy in work" a part of everyone's life. I offer *Quality Management Systems: A Practical Guide* as a monument to Dr. Deming's perseverance, tenacity, and unfaltering dedication to his mission.

ABOUT THE AUTHOR

Howard S. Gitlow, Ph.D., is Executive Director of the Institute for the Study of Quality and Professor of Management Science in the School of Business Administration, University of Miami, Coral Gables, Florida. He was a Visiting Professor at the Science University of Tokyo in 1990, where he studied with Dr. Noriaki Kano. His areas of expertise are quality management theory and statistical quality control.

Dr. Gitlow is a Senior Member of the American Society for Quality and a member of the American Statistical Association. He has consulted on quality, statistics, and related matters with many organizations, including several Fortune 500 companies. He is the co-author of several books including *The Deming Guide to Quality and Competitive Position* (1987, Prentice-Hall*); Quality Management: Tools and Methods for Improvement,* 2nd ed. (1995, Irwin); *Planning for Quality, Productivity, and Competitive Position* (1990, Dow Jones-Irwin); and *Stat City: Understanding Statistics through Realistic Applications,* 2nd ed. (1987, Irwin). Dr. Gitlow has also published numerous articles in the areas of quality, statistics, marketing, and management.

Dr. Gitlow received his doctorate from New York University. He is the recipient of numerous awards for outstanding teaching, outstanding writing, and outstanding research articles.

DEDICATION

To my mother,
Beatrice Alpert Gitlow
(1917–1999):
"You were the best. I love you."

CONTENTS

1

A QUALITY MANAGEMENT SYSTEM

PURPOSE OF THIS CHAPTER

The purpose of this chapter is to provide an overview of the "Quality Management System" presented in this book. The system combines Dr. W. Edwards Deming's theory of management and Japanese Total Quality Control into an integrated and internally consistent plan for the transformation of an organization.

DR. DEMING'S THEORY OF MANAGEMENT

Dr. W. Edwards Deming*

W. Edwards Deming was an American management consultant often called the father of the Quality Management movement. His teachings were considered a leading influence in the revival of the Japanese economy after Japan's defeat in World War II (1939–1945). In the 1980s, major corporations in the U.S. and other countries began to adopt his theory of management.

Deming developed a Quality Management theory that emphasized "joy in work." He said that quality should be stressed at each step of a process, not by inspecting the product or service once it is completed. In addition, Deming maintained that most product and service problems

---

* This section is abstracted from Gitlow, H. S., "W. Edwards Deming," *The World Book Encyclopedia*, World Book, Inc. (Chicago, IL), 1998, p. 119. For more information on Deming's personal history, see Kilian, C., *The World of W. Edwards Deming*, CEEPress Books (The George Washington University, Washington, D.C.), 1988.

result from faults in management, rather than from the carelessness of workers.

William Edwards Deming was born in Sioux City, Iowa. He received a doctorate in mathematical physics from Yale University in 1928. He was a mathematical physicist at the U.S. Department of Agriculture from 1927 to 1939. During World War II, he taught engineers how to use statistics to increase production of war supplies.

In 1950, a group of Japanese scientists and engineers invited Deming to Japan to lecture on the principles of quality control. Some Japanese companies that applied his methods increased their productivity and earned large profits. Deming's ideas spread. In addition to his work as a management consultant, Deming was a professor of statistics at New York University from 1946 to 1993.

Definition of Quality*

Quality is "a predictable degree of uniformity and dependability, at low cost and suited to the market."** For example, an individual buying a container of milk expects the milk to remain fresh at least until the expiration date stamped on the container, and wants to purchase it at the lowest possible price. If the milk spoils before the expiration date, the customer's expectation won't have been met, and he'll perceive the milk's quality as poor. Further, if this happens repeatedly, the customer will lose confidence in the milk provider's ability to supply fresh milk; in other words, the customer will feel he can't predict with a high degree of belief that the milk will be uniformly and dependably fresh.

Here's another example of quality: If an assembly line worker receives parts that are predictably dependable and uniform from the worker before her, her needs will be met and she'll perceive the quality of those parts as good. Similarly, if a hotel guest finds a clean, comfortable room containing all of the amenities promised, he'll feel that his expectations were met. But if the room isn't made up properly or lacks soap, the guest will perceive that the quality is poor.

---

* This section is taken from Gitlow, H., Oppenheim, A., and Oppenheim, R., *Quality Management: Tools and Methods for Improvement*, 2nd ed., Irwin (Burr Ridge, IL), 1995, p. 3. With permission.

** Paraphrased from Deming, W. E., *Quality, Productivity and Competitive Position* M.I.T. (Cambridge, MA), 1982, p. 229.

The Quality Environment*

Pursuit of quality requires that organizations optimize their system of interdependent stakeholders. This system includes employees, customers, investors, suppliers and subcontractors, regulators, and the community. The organization, which consists of employees and investors, must work together with suppliers and subcontractors to satisfy the needs of all stakeholders.

At one end of the system of interdependent stakeholders are an organization's external customers (its market segments). Each market segment's needs must be communicated to the organization through an ongoing process that conveys how an organization's products and services are performing in the marketplace and what improvements and innovations would optimize the system of interdependent stakeholders. The concept of *customer* also includes regulatory agencies, the community, and investors. The concept of customer should be applied to all areas and people within an organization. For example, customers are areas and people down the line.

At the other end of the system of interdependent stakeholders are the organization's suppliers and subcontractors. The organization communicates its customers' needs to its suppliers and subcontractors so that they can aid in the pursuit of quality for all stakeholders.

Employees are the most critical stakeholders of an organization. In the words of quality expert Kaoru Ishikawa,

> "In management, the first concern of the company is the happiness of people who are connected with it. If the people do not feel happy and cannot be made happy, that company does not deserve to exist.... The first order of business is to let the employees have adequate income. Their humanity must be respected, and they must be given an opportunity to enjoy their work and lead a happy life."**

Types of Quality***

Three types of quality are critical to the production of products and services with a predictable degree of uniformity and dependability, at low

---

* This section is taken from Gitlow, Oppenheim, and Oppenheim, *Quality Management,* pp. 3–4. With permission.
** Ishikawa, K. and Lu, D., *Total Quality Control? The Japanese Way,* Prentice-Hall (Englewood Cliffs, NJ), 1985, p. 97.
***This section is taken from Gitlow, Oppenheim, and Oppenheim, *Quality Management,* pp. 4–8. With permission.

cost, that are suited to the market. They are (1) quality of design or redesign; (2) quality of conformance; and (3) quality of performance.*

Quality of Design

Quality of design focuses on determining the quality characteristics of products or services that are suited to the needs of a market, at a given cost; that is, quality of design develops products from a customer orientation. Quality of design studies begins with consumer research, service call analysis, and sales call analysis, and leads to the determination of a product or service concept that meets the consumer's needs. Next, specifications are prepared for the product or service concept.

The process of developing a product or service concept involves establishing and nurturing an effective interface between all areas of an organization, for example, between marketing, service, and design engineering. Design engineering is one of marketing's customers, and vice versa.

Continuous, never-ending improvement and innovation of an organization's product and service concept require that consumer research and sales/service call analysis be an ongoing effort. *Consumer research* is a collection of procedures whose purpose is to understand the consumer's needs, both present and future. Consumer research procedures include both nonscientific and scientific studies. An example of consumer research is a study into the reasons why dog food purchasers buy or don't buy a particular brand of dog food. The investigation's goal is to determine the consumer's needs and redesign the package size, make the package reusable, or alter the dog food's composition. Consumer research should be ongoing so that the firm will always be in touch with changing consumer needs.

Consumer research can also be performed internally within an organization. For example, employees are the customers of some management policy decisions. Hence, employee surveys are a form of consumer research that could lead to improved management policy.

Sales call analysis involves the systematic collection and evaluation of information concerning present and future customer needs. Information is collected during sales interactions with customers. The analysis helps determine the customer's needs by examining the questions and concerns people express about products or services at the time of purchase. Sales call analysis is an important window into the customer's needs. An example of sales call analysis is a formal investigation into salesperson–customer

---

* The source for this material is Juran, J., *Quality Control Handbook*, 3rd ed. (McGraw-Hill (New York), 1979, pp. 2-4 through 2-9.

interactions at a personal computer distributorship. The investigation's purpose could be to collect information about the questions customers most frequently ask, and to use this data to improve the selling protocol.

Service call analysis is the systematic investigation of the problems customers/ users have with the product's or service's performance. Service call analysis provides an opportunity to understand which product features must be changed to surpass the customer's present and future needs. An example of service call analysis is Sony Corp.'s formal collection of information from field service technicians of customers' problems with Sony KV-32TW76 TV sets. The basic source document for the service call analysis data is the service ticket, which indicates the problem and the work done to solve it. This information is collected and over time may indicate problems that could require changes to work methods and/or materials, for example, redesigning the TV tuner or reducing the time between a customer's request for service and the completed service call.

Service call analysis can also be performed within an organization. For example, an area supervisor may examine the problems the next operation encounters using the parts/service/forms that his area delivers to the next operation. The purpose of the analysis could be to learn what the supervisor must do to pursue process improvement and innovation within his own area.

Quality of Conformance

Quality of conformance is the extent to which a firm and its suppliers can produce products and services with a predictable degree of uniformity and dependability, at a cost that is in keeping with the quality characteristics determined in a quality-of-design study. Once the specifications are determined via a quality-of-design study, the organization must continuously strive to surpass those specifications. The ultimate goal of process improvement and innovation efforts is to create products and services whose quality is so high that consumers (both external and internal) brag about them.

Some readers may question why the preceding paragraph states that specifications should be *surpassed*, rather than merely met. The rationale for this statement is that there's a loss associated with products that conform to specifications but deviate from the nominal or target value.

Quality of Performance

Quality-of-performance studies focus on determining how the quality characteristics determined in quality-of-design studies, and improved and innovated in quality-of-conformance studies, are performing in the marketplace. The major tools of quality-of-performance studies are consumer research and sales/service call analysis. These tools are used to study after-sales service, maintenance, reliability, and logistical support, as well as to determine why consumers do not purchase the company's products.

Quality and Cost*

Consumers can be grouped into market segments once the product characteristics (features) they desire are known and operationally defined. A characteristic is operationally defined if it is stated in terms of a criteria, a test, and a decision so that people can communicate with respect to its value. Features and price determine if a consumer will initially enter a market segment; hence, features and price determine market size. After the initial purchase, consumers' decisions to brag about a product or service, or purchase it again are based on their experience with the product or service, that is, the product's or service's dependable and uniform performance. Dependability and uniformity determine a product's or service's success within a market segment; therefore, dependability and uniformity determine market share within a market segment. Ultimately, features, dependability, uniformity, and price determine market size and market share.

Features

A loss in quality occurs when a process generates products or services whose features deviate from the needs of the individual (or group of individuals in a market segment), that is, when the products or services and/or price do not suit the market. This type of loss can be remedied by tailoring the product or service to the consumer's requirements and/or by modifying the product's or service's price. For example, shirt neck sizes may be marketed in tenths of an inch rather than in half inches, or Velcro® may be used in shirt collars instead of buttons. This segmentation strategy minimizes the loss in quality caused when the nominal levels of a product's or service's feature package deviate from the needs of an individual (or group of individuals) in a market segment, but it usually increases the cost of the product or service.

---

* This section is taken from Gitlow, Oppenheim, and Oppenheim, *Quality Management,* pp. 8–10. With permission.

Dependability and Uniformity

A loss in quality also results when a process generates products or services whose quality characteristics lack a predictable degree of uniformity and dependability (that is, when there is high unit-to-unit variation). High unit-to-unit variation results in rework and other problems that increase the cost of products or services, or conversely, greater uniformity results in lower costs of products or services. Lack of predictable uniformity and dependability in a product's or service's features will cause customers to lose confidence in that product or service. For example, if shirt neck sizes are manufactured to be 15 1/2 inches and customers notice variation from shirt to shirt, then this shirt-to-shirt variation will cause a loss in quality. This loss in quality can be reduced by understanding and resolving the causes of process variation. The two basic types of process variation — common and special variation — are discussed next.

Common and Special Variation

All systems (processes) vary over time. Consider a system such as your own appetite. Some days you're hungrier than usual, while other days you eat less than usual, and perhaps at different times. Your system varies from day to day to some degree. This is common variation. However, if you go on a diet or become ill, you might drastically alter your eating habits for a time. This would be a special cause of variation because it would have been caused by a change in the system. If you hadn't gone on a diet or become ill, your system would have continued on its former path of common variation.

Understanding the difference between common and special variation is critical to understanding W. Edwards Deming's theory of management. According to his theory, managers must realize that unless a change is made in the system (which only they can make), the system's capability will remain the same. This capability is determined by common variation, which is inherent in any system. Employees can't control a common cause of variation and shouldn't be held accountable for, or penalized for, its outcomes. Common variation can be caused by factors such as poor lighting, lack of ongoing job skills training, or poor product design. On the other hand, special variation can be caused by new raw materials, a broken die, or a new operator; it depends on the situation. Employees should become involved in creating and utilizing statistical methods so that common and special causes of variation can be differentiated, special variation can be resolved, and common variation can be reduced by management action. These actions will result in process improvements. Since unit-to-unit variation decreases the customer's ability to rely on a product's or service's dependability and uniformity, managers must understand how to reduce and control variation. Understanding and controlling variation leads to product or service improvement and innovation.*

Managers must balance the cost of having many market segments with the benefits of high consumer satisfaction caused by small deviations between an individual consumer's needs and the product characteristic package for his market segment. Also, managers must continually strive to reduce variation in product characteristics for all market segments.

The two sources of loss in quality must be detected in quality-of-performance studies. This information is then fed back into the quality-of-design studies and quality-of-conformance studies.

Quality and Productivity**

Why should organizations try to improve quality? If a firm wants to increase its profits, why not raise productivity? For years, W. Edwards Deming worked to change the thinking in organizations that operate with the philosophy that if productivity increases, profits will increase. The following example illustrates the folly of such thinking.

For the past 10 years the Universal Company has produced an average of 100 widgets per hour, 20% of which are defective. The board of directors now demands that top management increase productivity by 10%. The directive goes out to the employees, who are told that instead of producing 100 widgets per hour, the company must produce 110. Responsibility for producing more widgets falls on the employees, creating stress, frustration, and fear. They try to meet the new demands, but must cut corners to do so. Pressure to raise productivity creates a defect rate of 25% and only increases production to 104 units, yielding 78 good widgets, 2 less than the original 80.

Stressing productivity often has the opposite effect of what management desires. The following example demonstrates a new way of looking at productivity and quality.

The Dynamic Factory produces an average of 100 widgets per hour with 20% defective. Top management is continually trying to improve quality, thereby increasing productivity. Top management realizes that Dynamic is making 20% defectives, which translates into 20% of the total cost of production being spent to make bad units. If Dynamic's managers

---

* Gitlow, H. and Gitlow, S., *The Deming Guide to Quality and Competitive Position*, Prentice-Hall (Englewood Cliffs, NJ), 1987, pp. 9–10.
** This section is taken from Gitlow, Oppenheim, Oppenheim, *Quality Management*, pp. 13–15. With permission.

can improve the process, they can transfer resources from the production of defectives to the manufacture of additional good products. Management can improve the process by making some changes at no additional cost, so that only 10% of the output is defective on average. This results in an increase in productivity. Here management's ability to improve the process results in a decrease in defectives, yielding an increase in good units, quality, and productivity.

Benefits of Improving Quality

Deming's approach to the relationship between quality and productivity stresses improving quality to increase productivity. Several benefits result:

> Quality improves.
> Productivity rises.
> Cost per unit is decreased.
> Price can be cut.
> Workers aren't seen as the problem.

This last aspect leads to further benefits: less employee absence, less burnout, more interest in the job, and motivation to improve work.

In sum, stressing productivity means sacrificing quality and possibly decreasing output. Employee morale plunges, costs rise, customers are unhappy, and stockholders become concerned. On the other hand, stressing quality can produce all the desired results: less rework, greater productivity, lower unit cost, price flexibility, improved competitive position, increased demand, larger profits, more jobs, and more secure jobs. Customers get high quality at a low price, vendors get predictable long-term sources of business, and investors get profits. Everybody wins.

The 14 Points for Management

In 1980, Dr. Deming stated 14 interrelated points for management. Together, they provide a road map for leaders who want to transform their organizations.* These points are as follows.

---

* The 14 points are discussed in (1) Deming, W. E., *Out of the Crisis,* M.I.T. Center for Advanced Engineering Study (Cambridge, MA), 1986; (2) Gitlow, H. and Gitlow, S. *The Deming Guide to Quality and Competitive Position,* Prentice-Hall (Englewood Cliffs, NJ), 1987; (3) Scherkenbach, W., *The Deming Route to Quality and Productivity: Road Maps and Roadblocks,* Mercury Press (Rockville, MD), 1986; and (4) Neave, H., *The Deming Dimension,* SPC Press (Knoxville, TN), 1990.

1. Create constancy of purpose toward improvement of product and service, with the aim to become competitive and to stay in business and to provide jobs.
2. Adopt the new philosophy. We are in a new economic age. Western management must awaken to the challenge, must learn their responsibilities, and take on leadership for change.
3. Cease dependence on inspection to achieve quality. Eliminate the need for inspection on a mass basis by building quality into the product in the first place.
4. End the practice of awarding business on the basis of price tag. Instead, minimize total cost. Move toward a single supplier for any one item, with a long-term relationship of loyalty and trust.
5. Improve constantly and forever the system of production and service, to improve quality and productivity and thus constantly decrease costs.
6. Institute training on the job.
7. Institute leadership. The aim of leadership should be to help people and machines and gadgets to do a better job. Leadership of management is in need of overhaul, as well as leadership of production workers.
8. Drive out fear, so that everyone may work effectively for the company.
9. Break down barriers between departments. People in research, design, sales, and production must work as a team to foresee problems in production and in use that may be encountered with the product or service.
10. Eliminate slogans, exhortations, and targets for the work force that ask for zero defects and new levels of productivity.
11a. Eliminate work standards (quotas) on the factory floor. Substitute leadership.
11b. Eliminate management by objective. Eliminate management by numbers and numerical goals. Substitute leadership.
12. Remove barriers that rob the hourly worker of his right to pride of workmanship. The responsibility of supervisors must be changed from stressing sheer numbers to quality. Remove barriers that rob people in management and engineering of their right to pride of workmanship. This means, *inter alia,* abolishment of the annual merit rating and management by objective.
13. Encourage education and self-improvement for everyone.
14. Take action to accomplish the transformation.

Then, in 1989, Dr. Deming presented the theory underlying the 14 points for management; it is called "The System of Profound Knowledge."* In 1993, it was published in *The New Economics: For Industry, Government, Education*, M.I.T. Center for Advanced Engineering Study (Cambridge, MA), 1993, pp. 94–118. A second edition was published in 1994. The System of Profound Knowledge is discussed on pp. 92–115.

The System of Profound Knowledge**

The purpose of the System of Profound Knowledge is to transform leaders so that they will improve and innovate their organization's processes to promote *joy in work* for all of their stakeholders. Stakeholders include stockholders, employees, customers, suppliers, subcontractors, the community, the environment, regulators, and competitors, to name a few.

The System of Profound Knowledge comprises four components: appreciation of a system, theory of variation, theory of knowledge, and psychology. All four components are interdependent and no one component stands alone. Fortunately, it is not necessary to be expert in any of the components to understand and apply the System of Profound Knowledge. This discussion is not meant to be complete; its purpose is to present some of the highlights of Dr. Deming's theory of management.

Appreciation of a System

A system is a collection of components that interact and have a common purpose (aim). It is the job of top management to optimize the entire system toward its aim. It is the responsibility of the management of the

---

* Deming, W. E., "Foundation for Management of Quality in the Western World," paper presented at a meeting of the Institute of Management Sciences, Osaka, Japan, July 24, 1989. Deming, W. E., *The New Economics: For Industry, Government, Education*, M.I.T. Center for Advanced Engineering Study (Cambridge, MA), 1993, pp. 94–118. A second edition was published in 1994. The System of Profound Knowledge is discussed on pp. 92–115.

** This section of the chapter has been rewritten in the author's own words from Deming, W. E., *The New Economics: For Industry, Government, Education*, 2nd ed., M.I.T. Center for Advanced Engineering Study (Cambridge, MA), 1994, pp. 92–115. The author takes sole responsibility for any errors or omissions introduced due to his rewriting of the System of Profound Knowledge. Another description of the System of Profound Knowledge can be seen in Latzko, W. and Saunders, D., *Four Days with Dr. Deming: A Strategy for Modern Methods of Management*, Addison-Wesley (Reading, MA), 1995, pp. 33–44. Additionally, clarifications and insights on the System of Profound Knowledge can be seen on the Deming Electronic Network (DEN), a service of the W. Edwards Deming Institute, Washington, D.C.

components of the system to promote the aim of the entire system; this may require that they suboptimize their component.

Theory of Variation

Variation is inherent in all processes. There are two types of causes of variation: special causes and system causes. Special causes of variation are external to the system. It is the responsibility of local people and engineers to determine and resolve special causes of variation. System causes of variation are due to the inherent design and structure of the system; they define the system. It is the responsibility of management to isolate and reduce system causes of variation. A system that does not exhibit special causes of variation is stable; that is, it is a predictable system of variation. Its output is predictable in the near future.

There are two types of mistakes that can be made in the management of a system. First, treating a system cause of variation as a special cause of variation; this is by far the more common of the two mistakes — it is called *tampering* and will invariably increase the variability of a system. Second, treating a special cause of variation as a system cause of variation. Dr. Walter Shewhart developed the control chart to provide an economic rule for minimizing the loss from both types of mistakes.

Management requires knowledge about the interactions between the components of a system and its environment. Interactions can be positive or negative; they must be managed.

Theory of Knowledge

Information, no matter how speedy or complete, is not knowledge. Knowledge is indicated by the ability to predict future events with the risk of being wrong and the ability to explain past events without fail. Knowledge is developed by stating a theory, using the theory to predict a future outcome, comparing the observed outcome with the predicted outcome, and supporting, revising, or even abandoning the theory.

There is no true value of anything. Communication is possible when people share operational definitions. Operational definitions are statistical clarifications of the terms people use to communicate with each other. A term is operationally defined if the users of the term agree on a common definition.

Experience is of no value without the aid of theory. Theory allows people to understand and interpret experience. It allows people to ask questions and to learn.

Psychology

Psychology helps us understand people, the interactions between people, and the interactions between people and the system of which they are part. Management must understand the difference between intrinsic motivation and extrinsic motivation. *Intrinsic motivation* comes from the joy of a task well done; it comes from within an individual. *Extrinsic motivation* comes from fear of punishment or desire for reward; it comes from a source outside of an individual. All people require different degrees of intrinsic and extrinsic motivation. It is the job of a manager to learn the proper mix of the two types of motivation for each of his people.

PARADIGMS

A *paradigm* is a way of looking at the world. That way determines how an individual processes information and behaves. For example, one mother believes that a child should be constantly nurtured for him to develop into a happy and healthy person. She picks up her baby every time he cries at bedtime, even after she has assured herself nothing is wrong, such as a wet diaper or gas. Another mother believes that a child should be left to cry to develop a sense of independence. She lets her baby cry at bedtime, after having checked for common problems. Here you see two paradigms and their corresponding behaviors for dealing with the same set of facts, a crying baby.

Paradigm Shifts

Consider the following paradigm of pronunciation. Say the word "*ghoti*" out loud. Based on the traditional paradigm of pronunciation, you probably said "goaty, gotty, or hotty." There is another way to pronounce the word, based on a different paradigm of pronunciation. Pronounce the "gh" as in rough (f), the "o" as in women (i), and the "ti" as in nation (sh). Using the new paradigm, "ghoti" is pronounced "fish."

Accepting paradigm shifts is very difficult. If you overcome your need to rely on traditional models and change your paradigm of pronunciation, you will incorporate the new pronunciation into your daily speech. There will be serious consequences. People will not understand you. Some might even think you've lost your mind. You will be a lone soul, unless you can convince others to accept the new paradigm.

This is exactly what happens to those leaders who try to embrace the new paradigms for management, as described by Dr. Deming. Changing paradigms is a monumental task, because people cling to them for a sense of security.

Existing Paradigms

Leaders use paradigms, often without realizing that they are getting in the way of successful management. Some examples of existing paradigms are as follows:

1. Rewards and punishments bring out the best in people.
2. Focusing on results yields improvement of results.
3. Crisis intervention (fire fighting) will improve an organization in the long term.
4. Effective decisions can be made using "gut feel."
5. Rational decisions can be made using only visible figures.
6. Quality and quantity cannot be achieved together.
7. Winners and losers are necessary in most interactions.

Leaders who manage in the context of the above paradigms will be unable to survive in the 21st century. They don't know how to manage their organizations based on the new paradigms required for success in the current and future marketplace. Such leaders need a perspective from which they can understand and integrate the new paradigms of management.

New Paradigms

The System of Profound Knowledge explains four paradigm shifts that are necessary for organizational success in the 21st century. They are very different from traditional organizational paradigms.

Paradigm 1: Manage by improving processes to get results; do not manage just by demanding results.

Process and results management promotes improvement and innovation or organizational processes. Highly capable processes facilitate prediction of the near future, and consequently, a higher likelihood of achieving desired results. Results-only management causes people to abuse processes to get their desired results, and, ultimately, things get worse. Management by objectives, piece work, sales and production quotas, and time and motion standards are classic forms of results-only management. None of them improve the process that makes results. They just force individuals to do whatever must be done to get the desired results.

Process and results management is operationalized through the empowerment process. Empowerment* works at two levels. First, employees are empowered to develop and document best practice methods using the SDSA (Standardize-Do-Study-Act) cycle. Second, employees are empowered to improve and innovate best practice methods through application of the PDSA (Plan-Do-Study-Act) cycle.

SDSA Cycle. The SDSA cycle is used to define and document a process. It promotes standardization, which creates improved quality and reduced costs. The SDSA cycle includes the following four stages.

1. Standardize. All employees involved in a process (a work area) flowchart it from the perspective of how they actually perform the work. Next, they form a work team and agree on one best practice flowchart with appropriate key indicators. All employees recognize the need for one best practice method to ensure equal outcomes from the process under study.
2. Do. Team members test the best practice method as measured through key indicators in a planned experiment. The supervisor of the work team leads its members in a planned experiment to improve and innovate the process.
3. Study. Team members study the key indicator data from the planned experiment to determine the effectiveness of the best practice method. If the best practice method is lacking in some way, team members return to the "standardize" phase and develop a revised best practice method. If team members are happy with the best practice method, they proceed to the "Act" stage.
4. Act. Managers establish the standardized best practice method by putting it in the appropriate training manual and training all relevant employees.

PDSA Cycle. The PDSA cycle is used to improve and innovate a standardized best practice method. The PDSA cycle consists of four stages.

1. Plan. Employees involved with the best practice method form a work team to generate ideas for an enhanced best practice method. The enhanced best practice method (the Plan) is a revised version of the standardized best practice method flowchart in the current training manual; it includes key indicators.
2. Do. The Plan (revised flowchart) is tested on a small scale or trial basis as measured through key indicators in a planned experiment.

---

* Pietenpol and Gitlow, 1996.

3. Study. Team members study the key indicator data from the planned experiment to determine the effectiveness of the revised best practice method.

4. Act. Team members formalize the revised standardized best practice method by putting it in the appropriate training manual and training all relevant employees in it.

The PDSA cycle continues forever in an uphill progression of never-ending improvement.

Empowerment can only take place in an environment that nurtures and supports planned experimentation. Ideas for improvement and innovation can come from individuals or from teams (Plan), but tests of the worthiness of these ideas must be conducted through planned experiments under the auspices of the team (Do, Study, Act). Anything else will result in chaos, because everybody will do their own thing.

Paradigm 2: Manage by creating a balance between intrinsic and extrinsic motivation for each individual; do not rely only on extrinsic motivation to stimulate a person.

Intrinsic motivation comes from the sheer joy of doing an act, for example, the joy from a job well done. It releases human energy that can be focused on the improvement and innovation of a system. Intrinsic motivation cannot be given to an individual; it comes entirely from within the person experiencing it. Extrinsic motivation comes from desire for reward or fear of punishment, for example, the feelings stimulated by receiving a bonus. It comes from someone else, not the individual experiencing it. Frequently, extrinsic motivation can restrict the release of energy from intrinsic motivation by judging, policing, and destroying the individual.

A manager can create a fertile environment for others to experience intrinsic motivation in two ways. First, s/he can promote a workplace free of barriers between an individual and his or her ability to take joy in work. This can be accomplished through empowering employees as discussed in the prior paradigm shift. Second, s/he can hire and assign people into job positions that suit their personality and abilities. People are more likely to experience intrinsic motivation if they are performing a job for which they are suited.

Paradigm 3: Manage by promoting cooperation, not competition.

Conventional wisdom dictates that Americans are competitive people. Competition exists in family life, educational institutions, economic systems, and even in leisure games. In other words, competition is present in almost every aspect of American life. Americans believe in the value of competition due to four myths.* The myths are discussed below.

---

* Kohn, A., *No Contest: The Case against Competition*, Houghton Mifflin (Boston), 1986.

Myth 1: Competition is an unavoidable fact of life; it is part of human nature. Research shows that competition is a learned phenomenon. People are born neither cooperative nor competitive.

Myth 2: Competition motivates us to do our best. The evidence is overwhelmingly clear and consistent that competition almost never causes better performance. Superior performance not only does not require competition; it usually seems to require its absence. Competition never has to be present for skills to be mastered and displayed, or for goals to be set and met.

Myth 3: Competition is the best, if not the only way, to have a good time in contests or at play. Play is the opposite of work and has no goal other than pure enjoyment. Many play activities have come to resemble work because of competition. For example, amateur golfers play competitive golf, not just golf. They have become concerned with performance, not enjoyment. People need to play. The absence of play creates an unhealthy lifestyle for most people.

Myth 4: Competition builds character and increases self-confidence. A literature review concluded that cooperative learning situations, compared with competitive or individualistic situations, promote higher levels of self-esteem. The potential for loss is always present in competition; therefore, the more importance placed on winning in society, the more destructive losing will be. The costs resulting from competition are unknown and unknowable, but they are huge.

Is competition ever okay? Yes. If the aim of a system is to win, then competition is okay. However, in life, far fewer systems have winning as their aim than one might realize. For example, a woman plays racketball with a friend. What is the aim of the game? Is it to win or to have a good time and get exercise? Upon reflection, she realizes it is the latter, not the former.

Paradigm 4: Manage by optimizing the whole system, not the components of the system.

The whole system includes the interdependent system of stakeholders of an organization. Some stakeholders are investors, customers, employees, divisions, departments and areas within departments, suppliers, subcontractors, regulators, the community, and the environment.

Inter-system competition causes individuals, subsystems, or stakeholders to optimize their own efforts at the expense of other stakeholders. This form of optimization seriously erodes overall system performance. For example, investors demand a downsizing of employees in a year of record profit, or one department demands resources that it knows could be better used in another department.

Combinations of Paradigms: The above paradigms can be used individually, or in conjunction with one other as a system of paradigms to create new solutions to old problems.

Applying the System of Profound Knowledge

Each leader's concept of the new practice of management is based on the System of Profound Knowledge. However, even with that common theoretical base, leaders in an organization will have different interpretations of how they should practice management. Top management's task is to reduce individual-to-individual variation in respect to understanding the System of Profound Knowledge through education, training, and mentoring.

Since each organization is unique, with its own nuances and idiosyncrasies, managers in one organization cannot rely on the experiences of managers in other organizations to focus their transformation efforts. The conditions that led to the experiences of the managers in one organization may not exist for the managers in another organization.

The experiences of managers in one organization can stimulate the development of theories for improvement and innovation for managers of another organization. However, the leadership of each organization should develop its own model to operationalize the System of Profound Knowledge and the 14 points, integrating the "personality" of its organization.

JAPANESE TOTAL QUALITY CONTROL

After World War II, the economy of Japan was decimated. The leadership of Japan faced terrible economic and social crises. In the early 1950s, the Japanese Union of Scientists and Engineers (JUSE) invited Dr. W. Edwards Deming to speak to Japan's leading industrialists. At the time, he was the only person who offered any hope toward the resolution of Japan's problems.

Despite their reservations, Dr. Deming convinced the Japanese industrialists that by instituting his ideas, their quality could become the best in the world. Dr. Deming taught the Japanese the value of the Plan-Do-Study-Act (PDSA) cycle, the managerial significance of the distinction between special and system causes of variation, the value of statistical methods on the factory floor, that quality concepts are equally applicable in manufacturing and non-manufacturing environments, and to view an organization as a system (an interdependent system of stakeholders). The industrialists took Dr. Deming's teachings to heart and Japanese quality, productivity, and competitive position were improved and strengthened tremendously.

It is interesting to note that Dr. Deming's 14 points did not appear until 1980. What were the Japanese doing with respect to quality management between 1950 and 1980? The answer is that they* had taken the teachings of Deming and others and created their own school of thought on quality management, Japanese Total Quality Control (TQC).

Japanese TQC is empirically based on the experiences of what works in Japanese companies. It is unlike Dr. Deming's theory of management, which is theoretically based.** Furthermore, Japanese TQC has highly developed administrative systems by which organizational leaders can practice quality management. This is in contrast to Dr. Deming's theory of management, in which the leadership of each organization must develop its own administrative systems for quality management.

Japanese quality experts, with guidance from Dr. Joseph Juran and others, gave the world a great gift in the administrative systems of TQC. Administrative systems include education, training, and self-improvement; daily management; cross-functional management; and policy management. These systems are described in the following section.

A MODEL FOR QUALITY MANAGEMENT

Background

Top management, including the Board of Directors, initiates and leads quality management efforts. One of the first tasks for top management is to learn about the various theories, models, and techniques in the field. Then top management formulates a quality management model suited to the nuances of its organization. Quality management models will differ from organization to organization.

The Model

This book presents one possible model of quality management. Its theoretical base is Dr. Deming's System of Profound Knowledge and its practical base is the administrative systems of Japanese TQC. The administrative systems of Japanese TQC have been modified to be in harmony

---

* In 1949, JUSE formed the famous QC Research Group. Its members included Dr. Shigeru Mizuno, Dr. Kaoru Ishikawa, Mr. Masao Goto, Mr. Hidehiko Higashi, Dr. Tetsuichi Asaka, Dr. Masao Kogure, Mr. Shin Miura, and Mr. Eizo Watanabe.
** A comparison of Dr. Deming's theory of management and Japanese TQC can be found in Gitlow, H., "A Comparison of Japanese Total Quality Control and Dr. Deming's Theory of Management," *The American Statistician*, vol. 48, no. 3, Aug. 1994, pp. 197–203.

with the System of Profound Knowledge and the 14 points. This model is presented to stimulate the thinking of top managers when they develop a quality management model for their organization. It represents an "ideal" for promoting quality management that must be continuously pursued, and improved, by the leadership of an organization. The model presents a possible sequencing of activities that could be used to transform an organization (Figure 1.1).

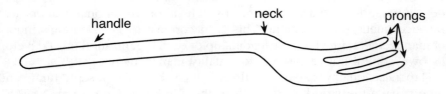

Figure 1.1 A Model for the Transformation of an Organization

The model is shaped like a fork with a handle, a neck, and three prongs. The handle is "management's commitment to transformation." The neck is "management's education." The first prong is "daily management." The second prong is "cross-functional management," and the third prong is "policy management." Each part of the fork is described below.

The fork analogy is used because quality management is an implement of nourishment for an organization. It feeds an organization, so its people have the energy to transform and pursue their goal of never-ending improvement.

The Handle: Management's Commitment to Transformation

Management's commitment to transform an organization is represented by the handle of the fork. The handle is the foundation for the neck and prongs and directs the action of the system. Without the handle, there is no fork. This is the case with quality management. Top management generates and directs the energy necessary to transform an organization. Top managers will expend this energy if they are confronted with a crisis or if they have a vision that they want to pursue.*

An example of management's commitment to transformation can be seen in the following situation. The Kelly family has experienced a crisis. Mr. Kelly lost his job and is unable to find a new one. Mr. and Mrs. Kelly

---

* Kano, N., "A Perspective on Quality Activities in American Firms," *California Management Review*, Spring 1993, pp. 14–15. See Quality Sweating Theory.

have to direct the family to transform its fiscal policies so that it can survive on one salary, instead of two. Their commitment to change comes from the impact of a crisis. On the other hand, the Lang family has a vision. Mr. and Mrs. Lang want to save money so that their children can attend college. They realize that saving at their current rate is not enough. They lead the family in changing its fiscal policies so that the focus is on saving more money. Their commitment to transformation comes from a long-term vision to provide the resources for the children's education.

Top management retains outside counsel to obtain expertise in the System of Profound Knowledge and to help management recognize its own strengths and weaknesses. The window of opportunity for the transformation opens. The outside expert assists top management in developing a plan for the transformation of the organization.

The top management then forms an Executive Committee (EC), which consists of all policy makers in the organization. The objective of the EC is to carry out the plan for transformation. This is accomplished by educating all members of the EC, as well as the members of the Board of Directors and stockholders (to the extent possible). The EC also develops a plan to communicate transformation activities to all relevant stakeholders.

Once the above phase of education and training is complete, the window of opportunity for transformation begins to close, unless the members of the EC exhibit concrete signs of transformation to relevant stakeholders. These signs include entering a period of education and self-improvement (neck), performing daily management (prong one), conducting cross-functional management (prong two), and performing policy management (prong three).

The Neck: Management's Education

The neck symbolizes the education and self-improvement activities of top managers. Studying the System of Profound Knowledge, the 14 points, and a transformation model is necessary to understand the theoretical and practical underpinnings of the new management. It will also provide insight to help cope with the upheaval that results from the transformation.

For example, the top management of an organization has made a commitment to transform their company through quality management. They hire a consultant to help them study the necessary theoretical information and integrate it into a model specific to their organization. The consultant also gives them feedback on their treatment of employees and the dynamics of how they interact with each other.

Mr. Harris, the CEO, is told that he is autocratic and uncompromising with employees; he needs to modify his behavior to be able to transform the organization. Ms. Gardner, the VP of Marketing, and Mr. Shepard, the VP of Production, compete with each other for ratings; they need to change so they can cooperate for the optimization of the system. The System of Profound Knowledge will help them understand the issues involved and define the behavioral changes that are necessary to proceed with the transformation.

Prong One: Daily Management

Prong one symbolizes the development, standardization, control, improvement, and innovation of methods (processes) used by employees in their daily routine. The development, standardization, and control of methods is called *housekeeping*. Housekeeping is accomplished through the Standardize-Do-Study-Act (SDSA) cycle. The improvement and innovation of "best practice" methods is called *daily management*. The reader is cautioned that daily management is used in two different contexts in this book. First, it describes developing, standardizing, deploying, maintaining, improving, and innovating the methods required for daily work. Second, it describes only the maintenance, improvement, and innovation of methods for daily work. Daily management is accomplished through the Plan-Do-Study-Act (PDSA) cycle. The PDSA cycle is a method that can aid management in improving and innovating processes. It guides management in reducing the difference between customers' needs and process performance. The PDSA cycle continues forever in an uphill progression of never-ending improvement.

An example of housekeeping is developing a "best practice" route (method) for driving to work. The day before you drive to a new job, you plan your initial route (Standardize). The next morning you actually drive the route in rush hour and experience the traffic patterns (Do). You consider your route based on the traffic patterns and information from people at work who drive the route (Study). The next morning, you alter your route (Act).

You will continue daily management, using the PDSA cycle until you come to an optimal "best practice" route. The "best practice" route will include contingencies depending on weather, road improvement work, and special events.

Prong Two: Cross-Functional Management

The purpose of cross-functional management is to develop, standardize, control, improve, and innovate organizational processes across divisions and departments. This is carried out to optimize quality, cost, delivery, service, quantity, and safety. Management carefully considers the effects of the optimization on sales and profits.

Cross-functional management includes the following activities:

1. Developing, standardizing, controlling, improving, and innovating cross-functional processes.
2. Developing measurements for cross-functional processes.
3. Coordinating and optimizing cross-functional processes with departmental processes.
4. Allocating resources for cross-functional and departmental processes by establishing targets.*
5. Ensuring that each department improves and innovates the cross-functional processes deployed to it in daily management.
6. Monitoring cross-functional processes in respect to targets from a corporate level.**
7. Utilizing the PDSA cycle to decrease the difference between process performance and customer requirements.

Finally, as expertise is developed with cross-functional processes, they are moved into daily management processes, if appropriate.

The EC manages and coordinates all cross-functional project teams. All cross-functional teams serve two functions. First, they serve as an opportunity for team members to acquire macro-level process knowledge and to study and learn Dr. Deming's theory of management. Second, they provide an opportunity for team members to identify and resolve process problems that cross department and division boundaries.

Blended families use cross-functional management to develop a viable family culture. When two people who have children from prior marriages form a family, it is difficult to coordinate and standardize the treatment of the children. The parents may have different goals, values, and processes in dealing with their respective youngsters. When the family is blended, it has to look for ways to define cross-familial issues and resolve them successfully.

For example, the Doles, a blended family, have an issue to deal with. Mrs. Dole has an 8-year-old daughter, Janis, who goes to bed at 9:00 P.M. Mr. Dole's 10-year-old son, Larry, has the same bedtime. Larry questions having the same bedtime as his 8-year-old stepsister. To resolve this issue, Mr. and Mrs. Dole have to discuss bedtimes and coordinate and improve

---

* Targets are used to allocate resources between processes.
** Targets are used to allocate resources between and within corporate and departmental processes.

their processes, taking into account the interaction between the former family units (departments).

Prong Three: Policy Management

Prong three represents policy management. Policy management is performed by using the PDSA cycle to improve and innovate the methods responsible for the difference between corporate performance and customer needs and wants, or to change the direction of an organization. Policy management assumes that housekeeping, daily management, and cross-functional management are functioning in the organization.

Policy management is accomplished through an interlocking system of committees. The Executive Committee (EC) is responsible for setting the strategic plan for the entire organization. It establishes values and beliefs, develops statements of vision and mission, and prepares a draft set of strategic objectives.

The Policy Deployment Committee (PDC) is responsible for deploying the strategic objectives throughout the entire organization. It develops an improvement plan (set of short-term tactics) for each department.

A Local Steering Team (LST) is responsible for implementing policy (short-term tactics) within a department, by coordinating and managing project teams. Project teams implement policy through improvement and innovation of the processes highlighted for attention. The management system is improved with each successive policy management cycle.

Policy management is used in family systems. Mr. and Mrs. Harris, EC of the Harris family, have a discussion. They would like to improve their family life with their children (mission). They decide that they should spend more time with the children (strategic objective).

They talk to their children about their objective to spend more time together (deploying the strategic objective). They propose eating dinner together on weekdays (short-term tactic). Each member of the family looks at his or her schedule, and they agree on the days that they are all free (implementing short-term tactics). They agree to try this system out for a week. Then they will meet and discuss how it might be improved in subsequent weeks.

Detailed View of the Model

A detailed view of the fork model is presented in Figure 1.2. It includes 34 steps, which are discussed in the remaining chapters of this book.

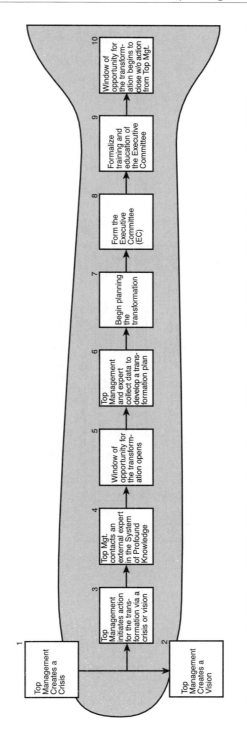

Figure 1.2 The Detailed Fork Model

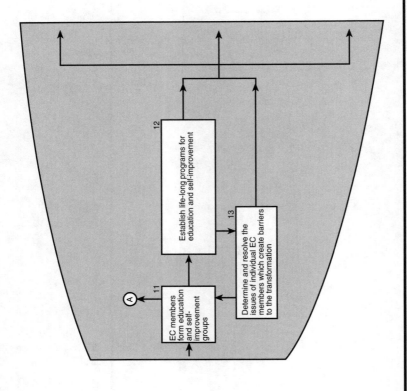

Figure 1.2 (continued) The Detailed Fork Model

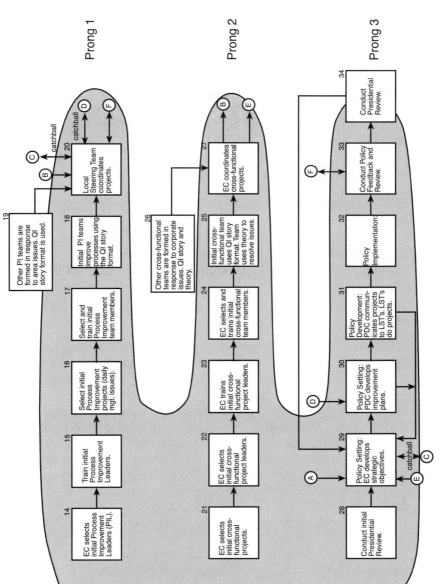

Figure 1.2 (continued) The Detailed Fork Model

SUMMARY

Chapter 1 presents an overview of the "Quality Management System" model that is the focus of this book. The model integrates Dr. W. Edwards Deming's theoretical "System of Profound Knowledge" and the empirical "Total Quality Control" of the Japanese.

Understanding the System of Profound Knowledge encourages leaders of organizations to give up existing ideas of management and adopt a perspective that embraces the new "win–win" paradigms: manage with a process and results orientation, manage to create a balance of intrinsic and extrinsic motivation, manage to promote cooperation, and manage to optimize the whole system.

Understanding Japanese TQC allows leaders to use highly developed administrative systems, based on the experiences of what works in Japanese companies. Administrative systems include education, self-improvement, daily management, cross-functional management, and policy management.

The model presented in this book is one possible quality management system. Its purpose is to stimulate leaders to develop their own models, based on the "personality" of their particular organization and the needs of their shareholders.

2

THE HANDLE: MANAGEMENT'S COMMITMENT TO TRANSFORMATION

PURPOSE OF THIS CHAPTER

Before any Quality Management efforts can be undertaken, the top management of an organization has to make a commitment to transformation. The purpose of this chapter is to explain what is required to sustain, coordinate, and promote that commitment.

STARTING QUALITY MANAGEMENT

Quality Management is a never-ending journey. However, all journeys begin with one step. The moment the leadership of an organization takes that first step, the organization has started Quality Management. The time required to complete the process described in this model depends on the resources allocated to the process.

The best time to begin Quality Management is now. Like a person who wants to lose weight and finds reasons not to start a diet, organizations manufacture excuses to put off the transformation. There is no specific time that is better than another to begin Quality Management.

Aids to Promoting Quality Management

Different needs and situations stimulate an organization to pursue Quality Management. Some examples of aids that promote the transformation of an organization to Quality Management include the desire to:

1. Exceed customer requirements.
2. Improve the organization's image.
3. Increase market size.
4. Improve employee morale.
5. Create a common mission.
6. Improve communication.
7. Standardize processes.
8. Create best practices.
9. Improve the physical environment.
10. Resolve problems before they become crises.
11. Bridge responsibility gaps.
12. Improve the documentation of processes, products, and services.
13. Improve the design of processes, products, and services.
14. Improve manufacturing and delivery of service.
15. Produce uniform products, at low cost and suited to the market (improve quality).
16. Increase profits.

Barriers to Quality Management

What stops an organization from pursuing quality? Examples of barriers that hinder the transformation of management of an organization include:

1. Inability to change the mind set (paradigms) of top management.
2. Inability to maintain momentum for the transformation.
3. Lack of uniform management style.
4. Lack of long-term corporate direction.
5. Inability to change the culture of the organization.
6. Lack of effective communication.
7. Lack of discipline required to transform.
8. Fear of scrutiny by supervisor.
9. Fear of process standardization.
10. Fear of loss of individualism.
11. Fear of rigidity.
12. Lack of financial and human resources.
13. Lack of training and education.
14. Lack of management commitment.

Top Management's Reluctance to Commit

Lack of management commitment will stop a Quality Management effort before it begins. If transformation promises improvement in all areas of the organization, why isn't it embraced by all top managers? One reason may be that many managers are unwilling to acknowledge company-wide success stories based on Quality Management theory.

Top managers may not be pro-Quality Management because it is not their own creation. On the other hand, they may fear failure to meet short-term goals or to manage effectively. Leaders are reluctant to change because they have been personally successful. The organization under them may be falling apart, but as long as they continue to get raises and positive performance appraisals, they can deny the rampant problems.

Leaders who verbally promote Quality Management but impede Quality Management by their actions create a situation called "the slow death." The slow death is similar to a plant whose leaves (workers), branches (supervisors), and trunk (middle management) have a natural inclination to grow, but the gardener (top management) does not provide water. Over time, the plant will die, as will Quality Management without the necessary nourishment of top management.

RESPONDING TO A CRISIS

Top management creates and directs the energy necessary to transform an organization. There are only two known sources for this energy, a crisis or a vision* (see steps 1 and 2 of the "Detailed Fork Model" in Figure 1.2).

Many companies begin a program of Quality Management as a reaction to crises discovered by top management. This section describes the crises in two companies, one Japanese and one American, which led to both embarking on Quality Management. They both successfully resolved their crises using Quality Management.

Juki Corporation

Juki Corp. is a Japanese manufacturer of products ranging from sewing machines to industrial robots. In 1973, Juki management uncovered external and internal crises, which led them to exert the energy necessary to make quality happen. The external crises included:

---

* Kano, N., "A Perspective on Quality Activities in American Firms," *California Management Review,* Spring 1993, pp. 14–15.

1. An inability to be competitive due to low quality and productivity.
2. Union problems.

The internal crises included:

1. Using the "genius approach" to research and development. Juki management relied on the creative abilities of employees to generate new products. This process did not allow management to predict, with any degree of accuracy and dependability, new improvements and innovations in products and services.
2. Behaving with a "market-out" point of view. Juki management created an organization in which products were produced and sold without determining the needs of customers.
3. Depending on the skill of individual workers to get the job done. Juki management relied on the nonreplenishable uniqueness of each individual to get jobs done, as opposed to standardizing work methods through training so that all relevant employees could do a particular job.
4. Acting as firefighters. Juki employees reacted to crises; they did not proactively improve processes to prevent crises from occurring in the future.

In response to these crises, Juki began a program of Quality Management. Juki Corp. challenged for, and won, the Deming Prize in 1976. The Deming Prize is the Quality Management equivalent of a black belt in karate. It is awarded by the Japanese Union of Scientists and Engineers (JUSE).

Florida Power & Light Company

Florida Power & Light Co. (FP&L) is the largest utility furnishing the generation, transmission, distribution, and sale of electricity in the State of Florida. It has experienced steady growth throughout its history. However, the pace of this growth increased dramatically between 1946 and 1947, making it difficult for FP&L's leaders to plan, finance, construct, and operate the utility. As FP&L grew, so did its managerial processes, becoming ever more cumbersome and unresponsive to customer needs. Nevertheless, because FP&L had been able to maintain stable prices for its customers it had avoided any potential crises.

In 1974, FP&L's ability to control costs was severely curtailed. In that year, OPEC's oil embargo and the subsequent increase in oil prices sent shock waves through the economy. Higher fuel prices quickly resulted in high inflation and declining sales growth. These external factors caused FP&L's stock price to fall as bond rates increased. Furthermore, in reaction to the oil crisis, the federal government passed the National Energy Act, which resulted in competition for utilities and promotion of conservation.

"By the early 1980s, FP&L was facing a hostile environment created largely by high inflation, decreasing customer sales, rising electric rates, and increasing fuel oil prices. The price of electricity was increasing faster than the Consumer Price Index (CPI)."* At the same time, competitive pressures were beginning to affect FP&L's long-term prospects. Customer dissatisfaction grew along with increasing expectations for reliability, safety, and customer service. In the meantime, FP&L's inability to react quickly to new environmental demands worsened its situation.

After developing a Quality Management process, FP&L challenged for, and won, the Deming Prize in November 1989. FP&L was the first non-Japanese company to win the prize.

Conclusion

In both of the above cases, top management uncovered crises that caused them to make a strong commitment to Quality Management and provide the leadership necessary to create quality.

CREATING A CRISIS

Top management can uncover and bring to the forefront the real or potential crises that face an organization (see step 1 of the Detailed Fork Model in Figure 1.2). One method top management can use to create a crisis is asking a probing question, such as, "What are the quality requirements of our major product/service demanded by our major customers?"** Frequently, top management is unable to answer this question. This may create a crisis for top managers, if they realize that they are out of touch with their customers' needs.

Another method by which top management can create a crisis is by conducting a brainstorming session on the crises that face their organization and analyzing the results with an affinity diagram, to be explained in the following sub-sections.

---

* From FP&L corporate document, "Description of Quality Improvement Program QIP-Corporate," Dec. 1988, p. 8.
** The author first heard Dr. Noriaki Kano of the Science University of Tokyo ask this question of executives in 1989.

Brainstorming Session

Brainstorming is a way to elicit a large number of ideas from a group in a short period of time. Members of the group use their collective thinking power to generate ideas and unrestrained thoughts. Brainstorming can be used to identify crises that face an organization.

Introduction

Effective brainstorming should take place in a structured session. The group should be small in number, generally between 3 and 12; having too large a group deters participation. Composition of the group should depend on the issue being examined. If the group is identifying crises facing the organization, it should include members from inside and outside the organization who come from extremely varied backgrounds with very different perspectives on the organization, and even on life in general.

The group leader should be experienced in brainstorming techniques. Seating should promote a free flow of ideas. A U-shaped or circle arrangement is suggested. The leader should record the ideas suggested by group members so everyone can see them, preferably on a flip chart, blackboard, or illuminated transparency.

Procedure

The recommended steps for a brainstorming session used to identify crises facing an organization are listed below. Prior to the brainstorming session, the following steps should be utilized. First, clearly state that the purpose of the brainstorming session is to determine "What crises face our organization?" Second, identify the individuals most appropriate for determining the crises facing the organization. Third, conduct a library search and/or Internet search on crises facing organizations in your industry. The results of this search are critical inputs to the brainstorming session to prevent "reinventing the wheel" brainstorming. Fourth, summarize the results of the library/Internet search and distribute them, in advance, to the members of the brainstorming session. In the distribution letter, ask group members to study and think about the information sent to them before attending the brainstorming session. Fifth, schedule the brainstorming session and take care of all the necessary logistics.

During the brainstorming session, the following steps should be utilized. First, each group member makes a list of crises on a slip of paper. This should take no more than 10 minutes. Second, each person reads one idea at a time from his or her list, sequentially, starting at the top of the list. As crises are read, they should be recorded and displayed by the group leader. Group members continue in this circular reading fashion until all crises on everyone's list are read. Third, if the next crisis on a member's list is a duplicate of a crisis already stated by another group member, that member goes on to the next crisis on his list. Fourth, members are free to pass on each go-around, but should be encouraged to add something, Fifth, the leader then requests each group member, in turn, to think of any new crises s/he hadn't thought of before. Hearing others' crises will probably result in group members thinking of new related crises. This is called *piggybacking*. The leader continues asking each group member for new crises, in turn, until the group can't think of any more. Finally, if the group reaches an impasse, the leader can ask for everyone's "wildest crisis." An unrealistic crisis can stimulate a valid idea about a crisis from someone else.

Rules

Certain rules should be observed by the participants in a successful brainstorming session — otherwise, participation may be inhibited. First, don't criticize, by word or gesture, anyone's crisis. Second, don't discuss crises during the session, except for clarification. Third, don't hesitate to suggest a crisis because it sounds silly. Many times a silly-sounding crisis can lead to a significant crisis. Fourth, only one crisis should be suggested at a time by each team member. Fifth, don't allow the group to be dominated by one or two people. Finally, don't let brainstorming turn into a gripe session.

Example

A brainstorming session was conducted at a private university during 1996 to identify crises. Prior to the brainstorming session, Internet and library searches were done by graduate students under the supervision of a professor with experience in such searches. The purpose of the searches was to identify crises at similar universities. The results of these searches were input into the brainstorming session.

The members of the brainstorming session included top-level managers and academics from selected divisions within the university. All members were viewed by senior management as being capable of identifying current crises and potential threats to the university.

A list of 174 crises was the outcome of that brainstorming session. A subset of the crises on that list is shown below.

1. Local funding going away
2. Federal funding drying up
3. State funding drying up
4. Grant money as a percentage of applications is declining
5. Competition for research awards is increasing
6. No mandatory retirement age
7. Some faculty are unable to get grants
8. The teaching load to keep all faculty busy may not be there
9. Shifting focus to learning rather than teaching
10. Business universities adding competition
11. Students focusing on "getting a job," not education for self-improvement
12. K–12 not doing its job
13. U.S. population is becoming more diversified
14. Drop in number of high school graduates
15. Majority of high school students less able to afford university education
16. Many schools have higher teaching loads (we look attractive to prospective faculty)
17. Some universities have lowered entrance qualifications
18. We are *not* focusing on the nontraditional students
19. Our schedule is set for the convenience of faculty and traditional students
20. Faculty are not student-focused
21. We need to focus on K–80 education
22. We must use our resources more effectively
23. Security costs are much higher
24. Security is vital to our image
25. The World Wide Web communicates image issues fast and worldwide
26. More money earmarked for federal regulations
27. Insurance rates have increased a lot
28. Need to beef up the use of technology
29. Need to use more interactive instruction
30. Need to educate staff and faculty that "Education is a business"
31. Use carrots to drive change – not the stick
32. Need to solicit ideas for change from the staff
33. We need to change our culture here
34. We don't work well together
35. We don't communicate among ourselves
36. We don't have the freedom to change gradually
37. We need to create or plan a crisis to get people moving
38. We can no longer afford to be all things to all people

39. We are good at finding crises, but not good at solving them
40. Academia has difficulty defining its stakeholders
41. It is easier to do it and apologize than to ask permission
42. University is a culture of asking permission
43. Innovative curriculum at some colleges within the university
44. Change is talked about, not financially supported
45. We allocate resources in ways that do not reward innovation
46. Funding is not based on strategic planning
47. Getting information technology is a crisis
48. Getting information technology used is a crisis
49. Difficult to change the culture and attitudes of people so they will use technology
50. We need better classrooms for our students
51. We need better dorm rooms for our students
52. Our dorm rooms need to be built with the future in mind
53. Professors view themselves as independent contractors, not as employees
54. Need to build trust with faculty
55. Security for access to scholarly communications is an issue
56. The Internet is changing the definition of publishing
57. Issues of "ownership" of information and charging for access need to be addressed
58. How scholars communicate is changing
59. Technology is changing the need for libraries in current format
60. We have a crisis in our social contract with society
61. People don't trust institutions any longer
62. Concerns about research ethics
63. People are "bitter" about their "contract" with the university
64. Some organizations and people are more "loyal" than others
65. We don't communicate the "depth" we have
66. We don't share or use the capability of one part of the university with other parts of the university
67. We need more entrepreneurial managers
68. Need to shift from "industrial age" to "information age" model
69. People need different skills at different times — JIT (just in time) training
70. Old attitude: "If we sit here — people will come to us"
71. We need more endowment
72. We need more dollars to use to change things
73. University needs to be more aggressive and open to money-generating innovations
74. What is the balance between our image and what we will sell?

Affinity Diagram

Once the brainstorming session yielded the 174 thoughts about crises, team members had to organize and consolidate them into a usable format. A tool that is very helpful in accomplishing this is the affinity diagram.

Introduction

An *affinity diagram* is used to organize and consolidate a large amount of verbal information. It fits the verbal data into natural clusters that bring out the underlying structure within the verbal data.

Procedure

Team members take the following steps to construct an affinity diagram. First, team members copy the brainstormed crises from the list constructed by the leader to 3 × 5 cards, one crisis per card. Second, the leader spreads the cards on a large surface in no particular order, making sure all cards face one direction. Third, all group members simultaneously move the cards into clusters so the cards in a cluster seem to be related. This step is performed in silence so that team members cannot influence each other. One team member may move a card into one cluster, and another team member may move the same card to another cluster; this may go on for some time, but the cards will eventually find a home cluster. Members continue to move cards until meaningful clusters emerge. Cards that do not fit into any cluster are placed into a miscellaneous cluster. Fourth, after members agree that clusters are complete (usually between 4 and 15 clusters emerge), the leader reads all the cards in a given cluster aloud and group members have a discussion to identify the theme of the cluster. Cards may change clusters in the process. The theme of a cluster is a short complete sentence that includes a noun and a verb. This procedure is performed for all clusters. The resulting set of clusters, each with its own header card, is called an affinity diagram.

Example

A subset of the affinity diagram developed from the 174 brainstormed thoughts about crises is shown below. The six major themes (crises) in the diagram, and their subthemes (crises), have been organized and reworded to enhance clarity and communicability to stakeholders of the university.

1. The university has an unclear understanding of its identity.
   Touchy-feely vs. virtual university
   Educators vs. trainers
2. The university has a decreasing resource base.
   Money is getting scarce
   Government funding is in short supply
   Difficulty cutting costs
3. The university has ineffective internal systems to keep up with tomorrow's demands.
   Need to improve infrastructure
   Need more and faster innovation
4. The university is encountering an increasingly hostile external environment.
   Increasing competition from business universities
   Increasing chaos and competition in medical services market
   Increasing demand to be accountable to stakeholders
   Need to broaden student base
   Technology is creating a new world
   Changing social contract with stakeholders
5. The university needs to improve the functionality of its culture.
   Faculty resists change
   Effectively use our capabilities with potential stakeholders
   Resource allocation system creates a negative environment
   Ineffective reward structure does not deal with technology
6. The university does not have an acceptable set of key indicators.
   Need to improve productivity
   Must have better results
   Need to improve stakeholder satisfaction

To summarize, the crises facing the university are an unclear understanding of its identity, a decreasing resource base, ineffective internal systems to keep up with tomorrow's demands, an increasingly hostile external environment, a culture in need of improvement, and an unacceptable set of key indicators.

CREATING A VISION

Top management can also initiate action for the transformation via a vision (see step 2 of the Detailed Fork Model in Figure 1.2). A vision can stimulate top management to expend the energy needed to transform an organization.

This idea is critical for organizations not facing a crisis. A vision can replace a crisis as a rallying point for the promotion of quality management. An important job of top management is to create a vision for the organization.

An example of a vision that drove top management to transform an organization is a situation that occurred in a social service agency. The agency, a group home program for troubled teenagers, was achieving its mission, adequately providing temporary shelter and basic care for adolescents separated from their families. However, the top management of the agency knew, through surveys of clients and referral agents, that the program needed to change to provide other services. These services included individual, group, and family therapy, academic counseling, and an overall plan coordinated by the clients, along with social workers, psychologists, house parents, teachers, and other involved staff members.

Top management had a vision of transforming the agency to one in which the needs of the clients were met in a more professional manner, utilizing a team to carry out an integrated plan. There was no crisis that stimulated this transformation. Top management saw a need to change the organization to exceed the clients' needs, which were not being addressed by the program in its current state.

One technique to create a vision that can be used by top management, or anyone for that matter, is to imagine the following scenario. The scenario requires that developer(s) of the vision personify the organization; that is, pretend the organization is a person.

> Imagine it is 100 years in the future and your organization has just died. All the stakeholders of the organization are standing around the coffin and the clergyman reads the eulogy. The eulogy ends with these words: Here lies insert the name of your organization, it was known and loved for insert the reason here.

The reason inserted above is the vision of your organization.

A *vision* should be a noble statement of long-term purpose. It should inspire people to take action to transform their organization.

Once top management has established a vision for an organization and its interdependent system of stakeholders, it can utilize brainstorming and the affinity diagram to identify issues that will prohibit realization of the vision. The topic of the brainstorming session can be, "What are the barriers that discourage realization of our vision?"

INITIATING ACTION FOR THE TRANSFORMATION

Top management initiates action for the transformation via a crisis and/or a vision (see step 3 of the Detailed Fork Model in Figure 1.2). Top management synthesizes, studies, and digests the crises facing the organization, as well as formulates and articulates the vision of the organization. If they feel it is warranted, they communicate the information about the crises and/or vision to relevant stakeholders. This process promotes commitment to the transformation among top management and stakeholders.

RETAINING OUTSIDE COUNSEL

After management has communicated the crises and the vision, the first action is retaining outside counsel (see step 4 of the Detailed Fork Model, Figure 1.2). Outside counsel is necessary for two reasons. First, expertise in the System of Profound Knowledge is not likely to be found within an organization. Second, organizations frequently cannot recognize their own deficiencies; that is, they don't know what they don't know.

It may not be easy to identify qualified outside counsel. Two methods are available that may help you to find and be able to afford outside counsel. First, you can read the literature and identify individuals who have written books and/or articles on Dr. Deming's theory of management and/or Japanese Total Quality Control. Many, but not all, of the available experts are prolific writers on the subject of Quality Management. Second, you can spearhead the formation of a Deming association in your area, if one does not already exist. A Deming association is a group of people wishing to improve their understanding of Dr. Deming's theory of management. Deming associations are called associations, study groups, research groups, user groups, forums, or alliances. They range from small, informal groups to large, highly structured organizations.

A list of Deming associations in the U.S. can be obtained from the Internet using the following URL: http://deming.ces.clemson.edu/pub/den/deming_assoc2.htm. A list of Deming associations in the U.K. can be obtained from the Internet using the following URL: http://deming.ces.clemson.edu/pub/den/deming_assoc5.htm. A list of Deming associations in countries other than the U.S. or the U.K. can be obtained from the Internet using the following URL: http://deming.ces.clemson.edu/pub/den/deming_assoc4.htm.

WINDOW OF OPPORTUNITY OPENS

Once outside counsel has been retained, a window of opportunity for the transformation opens (see step 5 of the Detailed Fork Model in Figure 1.2). The window of opportunity has an unspecified time limit that varies from organization to organization. If signs of transformation do not become obvious to the stakeholders of an organization, they will not believe that top management is serious about transformation, and the window of opportunity for transformation will begin to close. This is a common reason for the failure of Quality Management efforts in organizations.

COLLECTING DATA TO DEVELOP A TRANSFORMATION PLAN

An important role of outside counsel is to help top management assess the current status, and predict the future condition, of relevant stakeholders in respect to the transformation. They determine the "barriers against" and the "aids for" a fruitful transformation, at all levels within an organization and throughout the organization's interdependent system of stakeholders (see step 6 of the Detailed Fork Model in Figure 1.2).

Individuals have different reasons for wanting to, or not wanting to, promote Quality Management. Individuals will have different interpretations of what is involved in Quality Management. To be able to lead the Quality Management process, a leader must know each of his or her people's reasons for wanting, or not wanting, Quality Management, and how each of those different reasons interact with each other and with the aim of Quality Management. Consequently, a leader must obtain input from the stakeholders of his or her organization.

A *Gantt chart* is a tool for scheduling a plan. The following steps are required to construct a Gantt chart.

1. Identify all necessary actions for the plan.
2. Assign responsibility for each action required by the plan to an individual or department.
3. Determine the start and stop time for each action required by the plan.
4. Determine if any action items overlap. If the answer is yes, determine if there are enough resources available to simultaneously perform all overlapping actions.
5. List any comments for each action.

Figure 2.1 shows a Gantt chart for conducting a "barriers against" and "aid for" Quality Management plan. The top management appoints a team to conduct the analysis, asks outside counsel to conduct the analysis, or

Steps	Month 1	2	3	4	5	6	7	8	9	10	11	12	Comments
INTRODUCTORY STEP													
Develop a Gantt chart for the analysis.													
PLAN THE STUDY													
Identify real and potential crises.													See step 1 of the Fork Model.
Synthesize information about crises.													Top management studies and summarizes the real or potential crises facing the organization.
Write out real and potential crises.													Top management prepares a document that clearly describes the real or potential crises facing the organization.
Prepare a memorandum.													Top management prepares a memorandum explaining that a survey will be mailed to all employees which will study "barriers against" and "aid for" Quality Management. The memorandum contains the following informational items: (1) explanation of the crises facing the organization, (2) information on why Quality Management can help address the crises, (3) explanation that the output from the survey will be a series of action plans to deal with employees' concerns about Quality Management, and (4) guarantees concerning the anonymity of respondents.
Design the survey.													Team members design a survey that contains an informational section and a questionnaire section. The informational section includes information from the memorandum and instructions on how to complete and submit the survey. The questionnaire section contains the following questions: (1) In your opinion, what barriers will prevent Quality Management from working in your organization? (2) In your opinion, what aids will promote Quality Management in your organization?
Prepare a reminder message for non-respondents.													Team members prepare a reminder message for the non-respondents to the first distribution of the survey. The message should say: If you have already responded to this survey, please disregard this message.
Design the data collection plan.													Team members design a two-wave mail survey. The reminder message is mailed to the entire mailing list between the first and second waves to stimulate non-respondents to respond to the survey.
Distribute the memorandum.													See "Prepare the Memorandum" above.
COLLECT THE DATA													
Distribute the survey.													Team members distribute the survey to all employees. See "Design the Survey" above.
Collect the completed surveys.													Team members collect the completed surveys and determine the number of non-respondents.
Reduce number of non-respondents.													Team members recognize that non-response bias exists in a survey if non-respondents differ from respondents.
Send reminder message.													Team members send the reminder message to all persons who received the survey. See "Prepare a Reminder Message."
Determine the severity of non-response bias.													Team members determine if the responses to the second wave of the questionnaire are different from the first wave of the questionnaire. If they are similar, then it is assumed that non-response bias is not a severe problem, and first- and second-wave responses are combined and analyzed together. If they are different, then it is assumed that non-response bias is a problem and expert counsel should be asked for advice on how to rectify the situation.
ANALYZE THE DATA													

Figure 2.1 Gantt Chart for Conducting "Barriers Against" and "Aids For" Analysis

Steps	Month												Comments
	1	2	3	4	5	6	7	8	9	10	11	12	
Separate the questionnaire data.													Team members separate the data from question 1 and question 2. Question 1 yields "barriers against" data and question 2 yields "aids for" data.
Create a code book from the "barriers against" data.													Team members create a code book from the "barriers against" data. A code book is used to develop classification categories for verbal statements and to generate a frequency distribution of the number of verbal statements in each category.
Create a code book from the "aids for" data.													Team members create a code book for the "aids for" data.
Identify the root cause code book "barriers against" classification(s).													Team members consider the frequency counts for each classification and the effect a particular classification has on all other classifications when selecting the root cause classification(s) for which it is critical to develop action plans. Root causes can be identified with an interrelationship diagraph.*
Identify the root cause code book "aids for" classification(s).													Repeat the above for "aids for" classifications and frequency counts.
ACT ON THE ANALYSIS													
Identify action items for each "barriers against" and "aids for" root cause.													Team members determine the detailed action items necessary to resolve root cause "barriers against" Quality Management and to promote root cause "aids for" Quality Management.
Assign action items.													Team members, with the support of top management, assign action items to individuals or areas using a matrix diagram.* The rows of the matrix are action items and the columns of the matrix are people or areas. Team members study the matrix to create logical work loads.
Develop plans for each action item.													Responsible individuals or areas develop action plans to resolve "barriers against" Quality Management and/or to promote "aids for" Quality Management.
Approve action plans.													Top management approves all action plans or calls for their revision.
Initiate action plans.													Top management puts each action plan into play in the organization.
Check on the progress of action plans.													Team members periodically study the effect of the action plans in resolving "barriers against" Quality Management and in promoting "aids for" Quality Management.
Promote the action plans.													Top management promotes the action plans to create an environment favorable to Quality Management.

* Gitlow, H. and PMI, *Planning for Quality, Productivity and Competitive Position*, Dow Jones-Irwin (Homewood, IL), 1990.

Figure 2.1 (continued) Gantt Chart for Conducting "Barriers Against" and "Aids For" Analysis

some combination of the above two options. The start and stop times (see step 3 above) in each Gantt chart are a function of top management's urgency to transform to Quality Management.

Step 6 of the Detailed Fork Model in Figure 1.2 involves the "Introductory Step," the "Plan the Study Step," and the "Collect Data Step" of the Figure 2.1 Gantt chart. An explanation of each step appears in the far right column of the Gantt chart.

PLANNING THE TRANSFORMATION

Top management develops a transformation plan once the data has been collected in step 6 (see step 7 of the Detailed Fork Model in Figure 1.2). Step 7 involves the "Analyze the Data Step" and the "Act on the Analysis Step" of the Figure 2.1 Gantt chart. Again, an explanation of each step appears in the far right column of the Gantt chart.

FORMING THE EXECUTIVE COMMITTEE

The top management forms an Executive Committee (EC), which consists of all policy makers in the organization. The chairman of the EC is the President or Chief Executive Officer of the organization (see step 8 of the Detailed Fork Model in Figure 1.2). The EC should not exceed five or six members, plus a facilitator. It is important to include only policy makers on the EC.

TRAINING THE EXECUTIVE COMMITTEE AND BEYOND

The EC ensures that all of its members are trained in Dr. Deming's theory of management (see step 9 of the Detailed Fork Model in Figure 1.2). Training includes (1) introduction to Dr. Deming's theory of management (the System of Profound Knowledge, the 14 points, and the deadly diseases);* (2) psychology of the individual and team;** (3) basic quality control tools;*** and (4) administrative systems for quality.**** Training in

---

* An excellent text for this seminar is Deming, W. E., *The New Economics: For Industry, Government, Education*, 2nd ed., M.I.T. Center for Advanced Engineering Studies (Cambridge, MA), 1994.

** An excellent text for this seminar is Scholtes, P., *The Team Handbook: How to Use Teams to Improve Quality*, Joiner Associates (Madison, WI), 1988.

*** An excellent text for this seminar is Gitlow, H., Oppenheim, A., and Oppenheim, R., *Quality Management: Tools and Methods for Improvement*, 2nd ed., Irwin (Burr Ridge, IL), 1995.

**** This text is appropriate for this seminar.

the administrative systems for Quality Management includes developing competence in daily management, cross-functional management, and policy management. These last three forms of management will be discussed in Chapters 4, 5, and 6, respectively.

WINDOW OF OPPORTUNITY BEGINS TO CLOSE

Once the above phase of education and training is complete, the window of opportunity for the transformation begins to close unless the members of the EC take two actions (see step 10 of the Detailed Fork Model in Figure 1.2). First, they promote the plan to transform the organization (see the last line in the Figure 2.1 Gantt chart) from its current paradigm of management to a Quality Management paradigm. As stated earlier, the steps to develop and execute the plan are discussed in the rightmost column of the Figure 2.1 Gantt chart. Second, they diffuse Quality Management theory and practice within the organization and outside the organization to relevant stakeholders, for example, the Board of Directors, stockholders, suppliers, customers, regulators, and the community, to name a few.

Diffusion of Quality Management

The diffusion portion of this step of the model explains how to diffuse Quality Management among the different areas within an organization and from one organization to another organization (for example, suppliers, subcontractors, and regulators, to name a few).

How to diffuse innovations is not obvious. For example, creating a newsletter or having a meeting for all interested persons is *not* the way to reliably diffuse innovations. Other methods are needed. This section discusses such methods for both inter- (between) and intra- (within) firm diffusion.*

All potential adopters of Quality Management fall into one of five adopter categories: innovator, early adopter, early majority, late majority,

---

* The following references are the seminal works in the diffusion of information literature: Cool, K., Dierickx, I., and Szulanski, G., "Diffusion of Innovations Within Organizations: Electronic Switching in the Bell System, 1971–1982," *Organization Science*, vol. 8, no. 5, Sept./Oct. 1997, pp. 543–559; Rogers, E., *Diffusion of Innovations*, 1st, 3rd, and 4th eds., The Free Press (New York), 1962, 1983, and 1995; Rogers, E. and Shoemaker, F. F., *Communication of Innovations: A Cross Cultural Approach*, 2nd ed., The Free Press (New York), 1971.
Note: The Rogers and Shoemaker text is the second edition of *Diffusion of Innovations* under a different name.

and laggard.* Innovators are venturesome, cosmopolite, and friendly with a clique of innovators. They possess substantial financial resources and understand complex technical knowledge. However, they may not be respected by the members of their organization. They are considered to be unreliable by their near peers due to their attraction to new things. Innovators are frequently the gatekeepers of new ideas into their organization.

Early adopters are well respected by their peers; they are localite, opinion leaders, and role models for other members of their organization. They are the embodiment of successful, discrete use of ideas. Early adopters are the key to diffusing ideas such as Quality Management.

Early majority deliberate for some time before adopting new ideas and interact frequently with their peers. They are not opinion leaders. Late majority require peer pressure to adopt an innovation. They have limited economic resources that require the removal of uncertainty surrounding an innovation. Laggards are very localite and are near isolates in their organization. They are suspicious of innovation and their reference point is in the past.

Diffusion of Quality Management must consider several factors. First, it must involve opinion leaders in the diffusion of Quality Management. The EC identifies opinion leaders by asking themselves, "Who would we go to for advice about management within our organization?" They prepare a motivational plan to induce opinion leaders to try Quality Management. The motivational plan must have the commitment of the Executive Committee and should consider a balance of extrinsic and intrinsic motivators. Second, it must provide a Quality Management process that is adequately developed and not too costly for potential adopters at all levels within the organization. Third, it must develop the learning capacity of potential adopters of Quality Management. Fourth, it must systematically improve management's understanding of the factors that affect the success and/or failure of Quality Management and improve their ability to communicate said factors to potential adopters. Finally, it must increase intimacy between potential adopters and the diffusers of Quality Management.

If the above activities do not occur, or do not occur effectively, then the window of opportunity for the transformation to Quality Management begins to close. The next step to promote Quality Management is for the EC, with the assistance of outside counsel, to focus attention on top management's intellectual and emotional commitment to Quality Management. This occurs as the members of the EC enter the "neck" of the fork model.

---

* Rogers, *Diffusion*, 1995, pp. 263–265.

DECISION POINT

The end of the handle is the first critical decision point in the fork model. If the members of the EC discover that the energy to adopt Quality Management is not present in the organization, then a *"NO GO"* decision is made and all efforts toward Quality Management stop. On the other hand, if the members of the EC discover that the energy to adopt Quality Management is present in the organization, then a *"GO"* decision is made and the quality management effort proceeds to the neck stage of the fork model.

QUESTIONS FOR SELF-EXAMINATION

The following questions can be helpful in stimulating discussion in an organization that is considering a transformation to Quality Management.

1. Can Quality Management succeed without the commitment of your top management?
2. Is it necessary to accept all of the paradigms of Quality Management to start Quality Management? What are the paradigms?
3. Can your organization ease into Quality Management?
4. What are some barriers that hinder the transformation of your organization to Quality Management?
5. What are some aids that promote the transformation of your organization to Quality Management?
6. Does Quality Management apply to the service aspects of your organization?
7. Will our workers buy into Quality Management?
8. Will our unions buy into Quality Management?
9. How much training is needed for Quality Management, by level?
10. How much will Quality Management cost? Is it possible to compute this figure?
11. How long will it take to have Quality Management in our organization?
12. What is the best time to begin Quality Management?
13. Can another organization's Quality Management process become the blueprint for our organization's Quality Management process?
14. Is it helpful to visit organizations with successful Quality Management processes? If yes, why? If no, why?
15. What is being done in your organization to spread Quality Management?

SUMMARY

Chapter 2 presents a discussion of "the handle" of the fork model for Quality Management that is presented in this book. "The handle" is management's commitment to transformation, without which there can be no transformation. Aids to promoting Quality Management and barriers to it are presented. Lack of management commitment is a barrier that is addressed in this chapter.

Top management's reluctance to commit to Quality Management is discussed. The author postulates several reasons for it, including managers who are unwilling to acknowledge success stories of Quality Management, not pro-Quality because it is not their own creation, scared of failure to meet short-term goals or to manage effectively, and reluctant to change because they have been personally successful.

There are only two known sources for the energy needed by top management to transform an organization: a crisis or a vision. Two cases are presented that show how companies responding to crises were stimulated to begin a process of quality management. These companies are Juki (a Japanese manufacturer) and Florida Power & Light Co. (an electric utility).

If a company is not currently faced with an obvious crisis, top management can uncover and bring to the forefront any hidden crises that exist. This can be done by asking the question presented in Chapter 2, or by using brainstorming and the affinity diagram, also discussed in the chapter.

Another way in which top management can begin the transformation is by creating its own vision as a rallying point for the introduction of quality. This is critical for organizations that are not facing a crisis.

After top management makes the commitment to transformation, the first action is retaining outside counsel, because an expert in the System of Profound Knowledge will not likely be "in-house," and the organization frequently cannot recognize its own deficiencies. Outside counsel helps top management determine the "barriers against" and "aids for" transformation, and works with top management to develop a plan for the transformation.

Next, top management forms an Executive Committee (EC), which consists of all policy makers in the organization. The EC carries out the plan for transformation. Unless the members of the EC exhibit signs of transformation to relevant stakeholders, the window of opportunity for the transformation begins to close.

Questions for self-examination are presented at the end of Chapter 2, to stimulate thought and discussion in an organization that is contemplating transformation to Quality Management.

3

THE NECK: MANAGEMENT'S EDUCATION

PURPOSE OF THIS CHAPTER

After the top management of an organization commits to transformation, its members enter a period of education and self-improvement. The purpose of the neck of the Detailed Fork Model is to explain what top management needs to do to promote and coordinate its education and self-improvement in respect to transformation.

MANAGEMENT'S FEARS CONCERNING EDUCATION AND SELF-IMPROVEMENT

Education and self-improvement are both exciting and frightening processes. Top managers who have decided to become involved in these processes look forward to and fear them. They are anxious to learn about themselves and the improvement process, but they also have several concerns. The following are some questions they may be pondering:

1. What actually happens in meetings about Quality Management?
2. Will I lose power?
3. Will I be embarrassed?
4. Will I look stupid?
5. Will I "get it"?
6. Will I be able to do it?
7. Will I have to change my personality?
8. Will I be exposed as incompetent?
9. Will I have to justify myself to the others?

These fears and questions are a natural reaction to the task that lies ahead of top management. Education and self-improvement are difficult, soul-searching activities that have a profound effect on the individual and the organization. It takes courage and strength of character to involve oneself in these processes. The guidance of an outside expert and the support of colleagues who share the same concerns will be very valuable during this arduous process.

EDUCATION AND SELF-IMPROVEMENT GROUPS

One of the first tasks of the EC is forming one or more "education, training, and self-improvement" groups (see step 11 in the Detailed Fork Model in Figure 1.2). The aim of each group is to expand and deepen its understanding of Dr. Deming's theory of management in respect to business and life.

A group contains between three and six members, and meets frequently, for example, weekly. The areas of concentration are (1) studying the System of Profound Knowledge; (2) answering prescribed questions on Quality Management; (3) designating study teams for each of the 14 points; and (4) identifying and resolving personal barriers to transformation. The purpose of the above areas of concentration is to transform an individual's or organization's decision-making process from producing "lose–lose" or "win–lose" decisions to generating "win–win" decisions.

STUDYING THE SYSTEM OF PROFOUND KNOWLEDGE

The System of Profound Knowledge is discussed by the group, under the tutelage of an expert. The expert creates an environment in which group members deepen their understanding of how the System of Profound Knowledge might affect organizational and personal decision-making.

The expert may use group meetings, role-playing, case studies, or workshops to generate the individualized feedback necessary for each top manager so that s/he can transform personally, and consequently, promote the transformation of the organization.

The purpose of this type of session is not to "mess with" anybody's personal beliefs and values, but rather to make them aware of an alternative system of beliefs (the System of Profound Knowledge) and its potential impact on them and their decision-making processes.

Group Meeting and Role-Playing

An example of a group meeting in which the concepts of the System of Profound Knowledge are being discussed is presented below. The group consists of the QM Expert; the CEO of the company, Bob; the VP of Sales, Carol; the VP of Production, Ted; and the VP of Quality, Alice.

QM EXPERT:	Last time we were in the middle of a heated debate about intrinsic versus extrinsic motivation. I'd like to continue that discussion. Carol, you were having a hard time figuring out how you were going to motivate your people without sales incentives.
CAROL:	I still am. I've been thinking about it a lot. Some of my people will do a good job, even without the awards, but there are others who need that carrot to get them going.
TED:	I know what you mean. I feel the same way, except it's worse in Production. If I don't reward them for output, they'll just slack off.
QM EXPERT:	Alice, what are your feelings?
ALICE:	I agree with Dr. Deming's theory, but I'm at a loss as to what can replace our incentives. I understand Carol and Ted's concerns.
QM EXPERT:	Bob, what do you think about what you're hearing?
BOB:	I feel pretty much the same way, but I guess we have to try and look at it differently. We need to address everyone's concerns, and at the same time, we have to be looking for ways to change our system to be more in line with the direction we're going in.
QM EXPERT:	I feel that we're making some progress. Everyone seems open to some more information about alternative action. Last time you weren't as receptive. Let's get into how we can create joy in work at your company, without relying on extrinsic rewards. We'll start with each of you telling us what motivates you to do your job. Bob, let's hear from you first,
BOB:	Well, I have to tell you that I enjoy getting my paycheck. (*Everyone laughs.*) But that's an extrinsic motivator.
QM EXPERT:	I appreciate your honesty, Bob. Nobody expects you to work for no pay. But I'm sure there are other things that make you get up and come to work.
BOB:	Sure, I love what I do. It's a challenge running this company. I have to be on my toes, always thinking, planning, and learning. It's exciting watching my ideas get played out, especially when they work. I like

	mentoring people, watching them grow and develop. Do you want me to keep going?
QM EXPERT:	Actually, that's enough for now. You've given us a lot to work with. All the motivators that Bob mentioned, aside from his paycheck, of course, were intrinsic. Our task now is to transform the organization so that each and every employee in it can experience the same positive feelings about work that Bob has expressed. What I'd like to do now is some role-playing. Ted, I want you to pretend that you're one of the line workers in your department, Sam, reporting for his shift. Alice, I want you to be Ted.
TED (as SAM):	Good morning, Mr. Lawrence.
ALICE (as TED):	Hi, Sam. No time for chit-chat this morning. We've got to get to work to fill the Dynamic order that just came in.
TED (as SAM):	That might be a problem. The sorter was giving us a problem yesterday. Unless the night shift took care of it, it's going to have to be down for a while.
ALICE (as TED):	Why didn't you tell Jim yesterday?
TED (as SAM):	I tried to see Jim, but he was in a meeting with you the whole afternoon.
QM EXPERT:	Let's stop right there. Ted, what were you experiencing?
TED:	I was feeling bad and getting angry. First, the guy barely says good morning to me. Then he starts blaming me for something that's not my fault.
CAROL:	I think we all do that to the line workers. When we're under pressure, we take everything out on them.
BOB:	When you think about it, why would they want to come to work? No wonder we have such a high absentee rate.
QM EXPERT:	I think we're really onto something here. If we can make working at Universal a more positive experience for all employees, we can begin to create intrinsic motivation for everyone, not just Bob.

The above example demonstrates the role of the Quality Management (QM) expert and the desired atmosphere of the group setting. The QM expert is supportive, and provides continuity from one meeting to the next. S/he creates a nonjudgmental atmosphere in which group members are comfortable to express themselves. The QM expert includes everyone in the group in discussions and helps the leader become a role model for the others. The expert praises the group for growth and gently pushes the members when they are at an impasse. S/he summarizes where the group is going next and identifies the task in relation to the transformation.

Case Studies

Case studies in which top managers use the System of Profound Knowledge to change their decision-making process are shown below. The Quality Management expert works through selected case studies with each group. The studies the QM expert uses with a particular group are a function of the issues and concerns about Quality Management held by group members. It will take time and patience for top management to change their paradigms from the traditional set to the set proposed by Dr. Deming.

Business Example: The Drunk Employee

My name is Chuck. As the manager of a department in my company I supervise 27 employees. Carl, one of my subordinates, repeatedly comes to work drunk. His behavior causes productivity, safety, and morale problems among his co-workers. Additionally, his behavior affects the customers and suppliers of our department, the Human Resources department, his family and friends, and stockholders.

As part of my job, I try to understand the situation from the perspective of each stakeholder of my department. Whenever possible and appropriate, I talk to each stakeholder about the situation and identify their perspective.

> *Situation from Carl's perspective:* I'm in a lot of emotional pain. I don't know how to handle my situation. Everybody is on my back.
> *Situation from co-worker's perspective:* We're sorry that Carl developed a drinking problem, but why does the boss let him get away with it? He makes my job more dangerous. He could get me seriously hurt.
> *Situation from customer's perspective:* Carl's output isn't up to standard. He creates a lot of problems for me and for the people further down the line.
> *Situation from supplier's perspective:* My customers are starting to blame me for some of Carl's problems. I wish somebody would deal with Carl.
> *Situation from HR department's perspective:* We recently discovered Carl has developed a drinking problem that affects his performance on the job.

Situation from Carl's family and friends' perspective: We are worried about Carl. He is getting more emotionally distant and physically abusive every day. He needs help. We don't know what to do. We are worried that he will lose his job, then what will happen to us?

Situation from the stockholder's perspective: We want maximum profitability from the company, hopefully in the short term, but definitely in the long term.

Upon careful review of the situation from the perspective of each stakeholder, I conclude that there is no "win–win" solution, only "win–lose" or "lose–lose" solutions. Consequently, I think about the new paradigms presented in Chapter 1 to create options for resolving the situation such that all stakeholders can "win."

I study the new paradigms and decide that the paradigm "improve the process that creates results, don't just demand results" is an excellent new way to think. In other words, change the company's policies and procedures concerning drunk behavior and dealing with the drunk employee, don't just deal with the drunk employee. Given this shift in thinking, I develop the following potential "win–win" solution:

1. Organizational policies and procedures should be continuously studied and improved to decrease the frequency of employees who experience drinking problems. This is accomplished by improving hiring, training, and supervisory practices. This is a long-term solution.
2. The company must work with each employee who has a drinking problem. The employee gets help with his/her drinking problem. If he/she can't resolve his/her drinking problem in an Employee Assistance Program, the company will help him/her find a community program (e.g., Alcoholics Anonymous) that can help resolve drinking problems. While in the community program the company will provide benefits, but no compensation. The company will stick by the employee as long as the community program says the employee is working toward a positive resolution of his/her problem. The employee understands that if he/she stops making progress in the eyes of the community program, the company will terminate his/her employment.

This is a short-term solution for the company and the employee with the drinking problem, and a long-term solution for the company and all other employees who see the company's treatment of its sick employees.

Finally, I check to see if the solution using the new paradigm creates a "win–win" from the perspective of each stakeholder.

> *Chuck:* I believe that the solution is a "win–win" one. The company gets a more productive and joyful work force in the long run, and treats employees with drinking problems in a humane way in the short run. All employees see and understand both the short- and long-run behavior. The company realizes that it will have a superior work force in the long run and a more secure work force in the short run.
>
> *Carl:* I get help with my drinking problem. I believe that the solution is a "win–win." If I can't resolve my drinking problem in an Employee Assistance Program, the company will help me find a community program (e.g., Alcoholics Anonymous) that can help me. I understand that if I stop making progress in the eyes of the community program, the company will terminate my employment.
>
> *Carl's co-workers:* We believe that the solution is a "win–win" one. We like the policy that the company has adopted for dealing with employees with drinking problems. It is great that the company recognizes both its responsibilities and the worker's responsibilities in the policy.
>
> *Carl's customers and suppliers:* We believe that the solution is a "win–win" one. We like the policy that the company has adopted for dealing with employees with drinking problems. It is great that the company recognizes both its responsibilities and the worker's responsibilities in the policy.
>
> *HR department:* We believe that our policy creates a "win–win" solution for all stakeholders of the conflict.
>
> *Carl's family and friends:* We believe that the solution is a "win–win" one. Carl is getting help. The company is doing more than its fair share.
>
> *Stockholders:* We believe that the solution is a "win–win" one. An environment of employee well-being and concern is created that will yield the maximum return on our assets.

As we see, a true "win–win" solution has been created through adopting a new paradigm of management.

Company policies that deal with employee with drinking problems by providing time-limited substance abuse counseling create a possible wake-up call and consequent win for the employee and his/her family and friends. Chuck does not believe term counseling creates a "lose" situation for the employee with the drinking problem. Rather, it creates a needed wake-up call. Naturally, over time the need for post-termination substance abuse counseling will decrease due to improved hiring, training, and supervisory processes.

Daily Life Example: The Misbehaving Child

My name is Alice and I am a new mother. Recently, I brought my 1-year-old daughter, Lisa, to a restaurant for a nice lunch. I gave Lisa a plastic toy. She briefly chewed the toy and then threw it on the floor with great delight. I picked up the toy, scolded her not to throw it again, and gave her the toy. This scenario repeated itself 12 times over the course of lunch. Both my daughter and I got increasingly frustrated and upset playing this game.

As the decision maker, I identified the perspective of each stakeholder as best I could, given the situation.

> *Alice's perspective:* My child is misbehaving. It is embarrassing and I can't enjoy my lunch. Other patrons of the restaurant are looking at me.
>
> *Lisa's perspective:* I like throwing my toy on the floor, but I don't like mommy yelling at me.
>
> *Patrons' perspective:* I wish that mother would keep her child quiet.

I realized that a "win–win" solution did not exist given my current situation. Consequently, I considered the new paradigms for some possible insights. I realized that my method of inspecting my child's behavior was the relevant process in this situation. And, yelling at her was just demanding results that she couldn't deliver. So, I decided to change the process that makes results. My solution was to attach the toy with a string to my daughter's blouse. This way, she can throw the toy, but I don't have to pick it up. This process improvement clearly created a "win–win" for all stakeholders of the situation. I win because I can enjoy lunch. My daughter wins because she can throw her toy. The patrons win because they can enjoy a nice, quiet lunch.

Business Example: The Lean Organization

My name is Jan. I am a mid-level manager in my organization. The employees in my department are exhibiting signs of low morale. One of my jobs is to keep morale high to promote a productive work force. I am frustrated because there are few promotion slots and little money for pay raises that I can use to motivate my people. I don't know what to do other than to offer a kind word whenever possible.

I identified the other stakeholders of the situation. They are my boss, my employees, suppliers to my department, and customers of my department. I spoke with each of the stakeholders to identify their perspective on the situation at hand.

> *My boss' perspective:* One of the MBOs I set for my staff managers was to keep morale high in their departments. I said I would measure morale by comparing the average number of sick days per employee this year over last year. If the ratio is greater than 1.0 for one of my staff members, s/he is in trouble.
>
> *My employees' perspective:* Our manager is always pushing us for more productivity, but he never gives us more pay or bonuses. He always wants something for nothing. He never shares the company's wealth with us.
>
> *My suppliers' perspective:* Our customer (*my department*) is always trying to blame us for their quality problems. Why don't they clean up their act before they start blaming everyone else?
>
> *My customers' perspective:* The subcomponents we receive from our supplier (*my department*) have an unacceptably high proportion of defective parts. Also, the subcomponents frequently do not arrive on time.

Clearly, a "win–win" solution does not exist in the current situation. Consequently, I considered the new paradigms. I believe that promoting a balance of extrinsic and intrinsic motivation, as opposed to only extrinsic motivation, can be used to develop a "win–win" solution for all stakeholders. I decide that through empowerment (the SDSA and PDSA cycles), I will create an environment that will promote a work place in which employees can release their intrinsic motivation. I understand that each worker is simultaneously stimulated by intrinsic and extrinsic motivation. I recognize the constraints placed on my ability to manage due to limited extrinsic rewards. However, I am aware that I can create for my subordinates a work environment conducive to the release of their intrinsic motivation. I actively promote the empowerment process in my department.

Finally, I check with all stakeholders to determine if a "win–win" solution has been developed from their perspective.

Jan: I win because I am doing a great job; my employees' morale is high.

Jan's boss: My employees win because they experience joy in work and pride in the outcome.

My employees: We win because we enjoy our work.

My suppliers and customers: We win because we are working with a great partner.

A "win–win" situation has been created among the stakeholders of the original situation. As corporate conditions improve, top management should review its package of extrinsic motivators and consider options such as profit-sharing.

Daily Life Example: The Competitive Husband

My name is Martin. I have always enjoyed sports and am a very competitive person. I apply my competitive nature to all aspects of my life. My marriage is in trouble. My wife and I wife fight all the time. I don't like to lose arguments with my wife, or anybody else for that matter.

I tried to identify the problems in our marriage from my wife's perspective. She also thinks that our marriage is not doing well. She thinks I am very stubborn.

Clearly, my wife and I are deeply involved in a "lose–lose" situation. We can't think of anything to do to save our marriage. Consequently, we consider the new paradigms. We decide that we must start cooperating, instead of competing.

We agree to cooperate to find solutions to frequently occurring problems that create a win for both. Now, we keep a diary of arguments with each other and make a bar chart of the frequency of each type of argument. We select the most frequently occurring argument and brainstorm ways to eliminate it from our lives to create a "win–win" solution for both of us. Our marriage has been getting better over time. We both look forward to an improved relationship and future.

Business Example: The Profit Center

My name is Michael. I am the Chief Operating Officer of my organization. My company has five mills. I want to maximize the profitability of each mill. Nobody talks about it, but it seems that mill managers sometimes sabotage each other's profitability by demanding resources that could better be used elsewhere. An analysis shows considerable variation in mill manager rankings based on mill profitability. It seems to rotate, year to year.

The situation from the perspective of my mill managers seems to be that each one wants to be the number one mill in profitability, to get the largest bonus. Unfortunately, they will not share improved methods with each other. I cannot figure out how to make them cooperate to maximize overall profitability for the company. Consequently, I consider the new paradigms.

I decide that the problem may be that I am promoting optimization of the component parts of the system, not optimization of the entire system, through my compensation plan. I work with the mill managers to develop a compensation plan that is connected to the overall profitability of the organization (via profit sharing), as opposed to mill profitability.

Over time, my mill managers become a unified team. It is in each mill manager's best interest to help his fellow mill managers improve their operations. The entire top management team agrees that the new compensation creates a "win–win" for all stakeholders of the organization.

Conclusion

It is possible to change some "win–lose" and "lose–lose" situations within the context of existing paradigms into "win–win" situations within the context of new paradigms.

Workshops

Two examples of workshops that promote understanding of the System of Profound Knowledge are explained below. The first involves each member of the Executive Committee individually preparing an Executive Summary explaining the System of Profound Knowledge. The second involves the entire membership of the Executive Committee jointly developing a matrix explaining the relationships between the components of the System of Profound Knowledge and the 14 Points.

The first workshop involves the members of the Executive Committee individually preparing an Executive Summary for the QM expert explaining the System of Profound Knowledge in lay terms. The QM expert critiques each Executive Summary and returns it to its author. This process is repeated as many times as is necessary for each member of the EC to attain understanding of the System of Profound Knowledge, in the opinion of the QM expert.

The Executive Summary should contain (1) an introduction; (2) an explanation of the purpose of the System of Profound Knowledge; (3) a description of each of the underlying assumptions of the System of Profound Knowledge, with an example of each assumption; and (4) an explanation of each component of the System of Profound Knowledge, with examples of each relevant point.

Recall that the assumptions underlying the System of Profound Knowledge are (1) manage to improve the process that creates results; (2) manage using a balance of intrinsic and extrinsic motivation; (3) manage to promote cooperative efforts; and (4) manage to optimize the entire organization.

Also recall that the *some* of the salient points of the components of the System of Profound Knowledge are as follows. System theory involves defining a system, understanding who is responsible for stating the aim of the system, understanding the interdependence among the components of the system, and appreciating the need to optimize the entire system. The theory of variation involves understanding special and common causes of variation and knowing who is responsible for the resolution of each type of variation, understanding the meaning of stability and capability in respect to a process, and knowing when you can predict a process' future output. The theory of knowledge involves understanding how to acquire process knowledge, understanding that management is prediction, knowing how to develop operational definitions, and knowing why copying success can lead to disaster. Psychology involves understanding (1) "people, the interaction between people and circumstances"; (2) that "people learn in different ways, and at different speeds"; and (3) intrinsic and extrinsic motivation, and over-justification.

The second workshop requires that the entire Executive Committee prepare a matrix explaining the interrelationships between the four components of the System of Profound Knowledge and the 14 Points for Management. The rows of the matrix are the 14 points and the columns of the matrix are the four components of the System of Profound Knowledge. Each cell explains how a particular point (of the 14 Points) emanates from a particular component of Profound Knowledge. For example, part of the interrelationship between Systems Theory and Point 1 (Create constancy of purpose…) is that "an organization must have an aim stated by its management." Another example of an interrelationship is between the Theory of Variation and Point 1 (Create constancy of purpose…). It is that "management must work to reduce person-to-person variation of the understanding of the organizational aim."

In this workshop, the entire Executive Committee prepares a matrix for the QM expert. The QM expert critiques the matrix and returns it to the members of the Executive Committee. This process is repeated as many times as is necessary for the members of the Executive Committee to attain understanding of the System of Profound Knowledge, in the opinion of the QM expert.

ANSWERING PRESCRIBED QUESTIONS ON QUALITY MANAGEMENT

The group prepares, in writing, answers to an appropriate subset of the 66 questions in Chapter 5 of Dr. Deming's book, *Out of the Crisis.** It is important that group members reach consensus on the answers to each of the questions so they have a uniform understanding of what quality means within their organization and so they learn participatory management skills.

Group members follow the PDSA cycle in this endeavor, and they spend at least 2 hours per week discussing each question. One hour is devoted to developing a plan (Plan) for collecting the data required to answer the question and establishing a team to collect said data. Next, the plan is carried out (Do) by group members. The results are studied (Study) by group members, who then spend at least 1 more hour discussing the answer to each question. The group facilitator writes up the consensual answer to each question (Act).

The written answer is sent to an external expert in Dr. Deming's theory of management so that s/he may use it as a basis for guiding the group's education and self-improvement process. That is, the expert will use the written answer to help group members establish an improved Plan for understanding the question under study. The 2 hours of study per week per question mentioned earlier represents an arbitrary amount of study time and should be modified in accordance with the progress of the group. The period of education and self-improvement should be extended indefinitely into the future (see step 12 of the Detailed Fork Model in Figure 1.2).

The following section is an example of correspondence between a Quality Management expert and members of an organization working on transformation. First, the initial letter sent by the outside counsel to begin the process of answering the prescribed questions is presented.

---

* The method for educating top management was pointed out to the author by William Latzko, Bergen County, New Jersey. The 66 questions for self-examination can be found in Deming, W. E., *Out of the Crisis,* M.I.T. Center for Advanced Engineering Studies (Cambridge, MA), 1986, pp. 156–166.

To: Bob Walters
        Chief Executive Officer
        Universal Company
From: Quality Management Expert

Dear Bob:

I am glad to hear that Universal is going to begin the never-ending journey toward quality. The journey is long and difficult. Education is your guide.

Your organization's education in the area of the improvement of quality should begin with top management, you and your immediate staff reports. The more top management studies the philosophy and techniques for the improvement of quality, the quicker Universal will progress on its transformation to Quality Management.

Initially, this education will require that top management meet weekly to discuss and prepare in writing answers to the 66 questions in Chapter 5 of Dr. Deming's book, *Out of the Crisis*. It is important that top management come to consensus in respect to the answers to each of the 66 questions so they have a uniform understanding of what quality means at Universal and so they improve their participatory management skills.

Top management should plan to spend at least 2 hours per week discussing each question, using the PDSA cycle in this endeavor. One hour is devoted to developing a plan (Plan) for collecting the data required to answer the question and establishing a team to collect said data. Next, the plan is carried out (Do) by team members. The results are studied (Study) by top management. Management spends at least 1 more hour discussing the answer to each question. The Executive Committee facilitator writes up the consensus answer to each question (Act).

The written answer is sent to me so that I may use it as a basis for guiding management's education at the monthly meeting of the Executive Committee. I will use the written answer to help management establish an improved Plan for understanding the question under study.

The 2 hours of study per week per question mentioned earlier represents an arbitrary amount of study time and will be modified in accordance with the educational progress of EC members.

If you have any questions, give me a call.

The following letter is the expert's response to the company's answers to the first of Dr. Deming's 66 questions in *Out of the Crisis*. Question 1 appears below for your convenience.

1a. Has your company established constancy of purpose?
1b. If yes, what is the purpose? If no, what are the obstacles?

1c. Will this stated purpose stay fixed, or will it change as presidents come and go?
1d. Do all employees in your company know about this stated constancy of purpose (*raison d'etre*), if you have formulated one?
1e. How many believe it to the extent that it affects their work?
1f. Whom does your president answer to? Whom do your board of directors answer to?

To: Bob Walters
         Chief Executive Officer
         Universal Company
From: Quality Expert

Dear Bob:

Congratulations on taking a significant step toward the pursuit of quality. Please explain to the Executive Committee the importance of not becoming frustrated by my extensive critique. My critique is not a judgment of their work; rather, it is an attempt to improve their understanding of Dr. Deming's theory of management. The answer–critique–answer cycle is the method by which they will learn how to pursue Quality Management.

Please remind them that the point of answering the questions and the critiques is *education*, not to get through all 66 questions. To that end, please find my critique in the form of questions and comments for your Executive Committee's written consensus answer to question 1 in Chapter 5 of Dr. Deming's book, *Out of the Crisis*.

General Comments (Prelude to Question 1)

Did your committee follow the PDSA cycle when answering question 1? Or did they just have a discussion (Plan) and answer writing session (Act), with no Do and Study?

1. Did they develop a plan (Plan) for collecting the data required to answer question 1 and establish a team to collect said data?
   a. If yes, where is the data?
   b. If no, why not?
2. Did the duly appointed team carry out (Do) the Plan?
3. Did the duly appointed team study the results (Study) and report back to the Executive Committee?
   a. If the study phase yielded negative information, did the Executive Committee punish the team for bringing back unfavorable results?

b. If the study phase yielded positive information, did the Executive Committee modify the organization's mission (Act)?

Comments on Question 1a

Has your company established constancy of purpose?

Your answer: "No."

You must not be frustrated by not having constancy of purpose after all of your work in this regard to date. The critical point is to build a clear vision, mission, set of values and beliefs, and strategic plan with tactics out of your current vision and mission statements, quality policy, company values, and strategic plan. Your problem is that you have not integrated the information you have. This causes conflict and confusion. This conflict and confusion will grow greater as you move down in the organization due to interpretation and reinterpretation of the mission.

Universal's vision is an ideal state that employees can pursue, even though they may never get there. The mission is the vehicle by which employees will pursue their vision. The values and beliefs set the boundaries on employees' behavior in their pursuit of the vision via the mission. Finally, the mission, in conjunction with your values and beliefs, sets the stage for management to establish strategies and tactics.

Comments on Question 1b

If yes, what is the purpose?

If no, what are the obstacles?

Your answer: The obstacles include:

1. "We have communicated too many statements to employees... Although these statements complement each other, there is no specific statement of constancy of purpose."
2. "...our actions do not support our words."

First, let me address the obstacle of communicating too many quality-related statements to employees, without a statement of constancy of purpose. The Executive Committee's resolve to this obstacle is that constancy of purpose will be achieved by communicating Universal's values and beliefs to all employees. In my opinion, a better approach would be to integrate your vision, mission, values and beliefs, strategies, and tactics so that they provide a clear road map to all employees when they are deployed throughout the organization.

To do this, the Executive Committee forms a cross-functional team. The basic question to be addressed by the team is "What are the obstacles that prevent Universal from pursuing its vision and mission?" The team considers barriers to integrating the vision, mission, values and beliefs, strategies, and tactics when answering the above question.

Second, let me address the obstacle of management's actions not supporting management's words. Point 1 of Dr. Deming's 14 points addresses exactly this issue. Establishing constancy of purpose, through an integrated vision, mission, set of values, strategies, and tactics, is equivalent to establishing a desired nominal value for a process. Getting all employees to uniformly interpret and follow said purpose is a problem of the reduction of variation.

One important job of the Executive Committee members is to manage, and get their people to manage, in respect to Universal's purpose. How you go about doing this is covered in Dr. Deming's points on training, supervision, and barriers to pride of workmanship. It is important that the quality expert keep working with the top management until they understand and exhibit the System of Profound Knowledge in the conduct of their daily work. When top management has made the transformation, the quality expert provides education and training throughout the rest of the organization.

Comments on Question 1c

Will this stated purpose stay fixed or will it change as presidents come and go?

Your answer: "By using Universal's values and beliefs, the purpose will not change."

The answer to question 1c is really unknown. Only a succession of presidents and time will tell the real story. However, the improvement process continues, assuming the answer to question 1c is "stay fixed."

Comments on Question 1d

Do all employees of your company know about this stated constancy of purpose, if you have formulated one?

Your answer: "All employees are familiar with Universal's values and beliefs. They probably do not know about our constancy of purpose."

How do you know that the employees are familiar with the company's values? Where is the data? It is important that you manage by theory and facts, not by unsubstantiated opinion and guesswork.

How to communicate constancy of purpose throughout your organization is not obvious, for example, creating a newsletter or having a meeting for all interested persons is *not* the way to reliably communicate constancy of purpose. Spreading constancy of purpose includes several factors. First, it must involve opinion leaders within the various departments and areas of an organization. Second, it must provide a worthy and noble "purpose." Third, it must consider the learning capacities of potential adopters. Fourth, it must systematically improve management's ability to communicate the "purpose" in words and deeds. Finally, it must create intimacy between the "purpose" and stakeholders of the organization.

Comments on Question 1e

How many believe it to the extent that it affects their work?

Your answer: "...a small number..."

How do you know that the company's purpose affects only a small number of employees in the way they perform their jobs? Where is the data? In what ways does the company purpose affect the way employees do their jobs? Are these effects positive or negative? Do you have any data to explain these effects? What are the characteristics of employees who are affected? Not affected? What explains the difference? This data may help the Executive Committee develop a plan to spread Universal Company's purpose to all employees.

Comments on Question 1f

Whom does your president answer to?

Whom does your board of directors answer to?

Your answer: "Customers and employees."

I agree. Ultimately, the president and the board of directors answer to customers and employees. If your customers do not experience joy with your products and services, which are the result of the decision-making process of the president and the board, they will not give Universal the gift of business. Also, if your employees do not take pride in their work and joy in the outcome, they will not give Universal the energy required for a successful business. Clearly, this affects profit, return on investment, and dividends. Stockholders experience joy from the same set of managerial decisions.

The following discussion may shed some new light on your answer to question 1f. The old way of thinking about an organization is to visualize it as a pyramid in which decisions move from top management, through middle management, to line workers, and eventually to customers. In this type of pyramid, everyone's effort is focused on making the person(s) higher up in the pyramid (their boss) happy. This is fueled by administrative systems such as performance appraisal. Please note that making your boss happy is very different from making your customer happy.

Top

Middle

Line

Customers

A new way of thinking about an organization is to visualize it as an inverted pyramid in which decisions move from the base of customers, through line workers, through middle management, to top management, and then back up through the pyramid to customers. In this type of pyramid, everyone's effort is focused on making the person(s) higher up in the pyramid (their customer) happy. The president's job is to support top management, top management's job is to support middle management, and so on, throughout the organization. Making customers and employees happy is what your president and board should be doing.

Customers

Line

Middle

Top

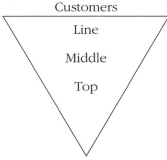

I hope my critique is helpful. If you have any questions, give me a call.

DESIGNATING STUDY TEAMS FOR EACH OF THE 14 POINTS

The members of the EC create study teams for each of the 14 points. The job of each study team is to remove organizational barriers to the transformation in respect to its point, by studying and using the System of Profound Knowledge. The activities of each team may lead to the establishment of a cross-functional team (see step 21 of the Detailed Fork Model in Figure 1.2), an educational program for members of the EC (see step 12 of the Detailed Fork Model in Figure 1.2), or input into the policy management process (see step 29 of the Detailed Fork Model in Figure 1.2).

An example of a study team is one that is created to study point 12, "Remove barriers that rob employees of their pride of workmanship."* One of the barriers that is identified by the group is the company's performance appraisal system. The cross-functional team uses the System of Profound Knowledge as the basis for discussing, redesigning, and continuously improving the performance appraisal process. The case study of Field of Flowers, presented in Chapter 5, offers further information on study teams using the System of Profound Knowledge to redesign and improve systems that create barriers to transformation.

IDENTIFYING AND RESOLVING PERSONAL BARRIERS TO TRANSFORMATION

Each member of the EC identifies and resolves his or her issues that create barriers to the transformation (see step 13 of the Detailed Fork Model in Figure 1.2). Examples of individual barriers to the transformation are the following beliefs:

1. Extrinsic motivators bring out the best in people.
2. Focusing on results yields improvement of results.
3. Fire fighting will improve an organization in the long term.
4. Effective decisions can be made using guesswork and opinion.
5. Rational decision-making can be performed using only visible figures.
6. Quantity is inversely related to quality.
7. Your most important customer is your superior.
8. Competition is superior to cooperation, and winners and losers are necessary in any interaction.

Each member of the EC examines his or her own opinions to determine if he or she holds any of the above beliefs or others that would impede the transformation. The quality management expert can use the following questionnaire to pinpoint problem areas that individuals may be experiencing.**

---

* Deming, W. E., *Out of the Crisis*, p. 77.
** McNary, L., *The Deming Management Theory: A Managerial Leadership Profile for the New Economic Age*, Ph.D. dissertation (May 1993), The University of New Mexico (Albuquerque, NM). On microfilm with the University of Michigan (Ann Arbor, MI).

Directions:

1. Below is a list of managerial traits (labeled A through J) that have been placed on a 1–10 scale.
2. Place an "X" on each scale to indicate the degree to which you exhibit a particular trait in your daily management style.*
3. Remember, there are no right or wrong answers.

A. With respect to employees, I believe in:

1	2	3	4	5	6	7	8	9	10

Fostering motivation through rewards and punishment   Neutral   Fostering motivation through pride and satisfaction in work

B. With respect to employees, I believe in:

1	2	3	4	5	6	7	8	9	10

Managing for the short term using daily production reports and/or quarterly measures of profit   Neutral   Managing for the long term by expanding the market for all

C. In making any organizational decision, I believe in:

1	2	3	4	5	6	7	8	9	10

Focusing on the result of a process with an emphasis on numerical goals and quotas   Neutral   Focusing on the process to get results with an emphasis on improvement and innovation

---

* In Dr. McNary's dissertation this instruction reads as follows: "Place an 'X' on each scale to indicate the degree to which *you possess* a particular trait."

D. I believe in:

1	2	3	4	5	6	7	8	9	10

Managing with a
tradition
hierarchial   Neutral
organizational
chart

Managing with
an emphasis on
customer–
supplier
relationships

E. For the success of my organization, I believe that:

1	2	3	4	5	6	7	8	9	10

Optimizing my
own situation
or competition   Neutral
is most
important

Optimizing the
system of
interdependent
components or
cooperation is
most important

F. With respect to employees, I believe in:

1	2	3	4	5	6	7	8	9	10

Managing by
formal appraisal   Neutral
process

Managing
through
informal
feedback and
coaching

G. I believe in:

1	2	3	4	5	6	7	8	9	10

Basing decisions
on visible
figures such as   Neutral
ROI calculations

Basing decisions
on operational
definitions,
visible figures,
and considering
the effect of
invisible figures

H. I believe that:

1	2	3	4	5	6	7	8	9	10

Management cannot be responsible for all components of an organization

Neutral

Management is responsible for all components of an organization

I. I believe that:

1	2	3	4	5	6	7	8	9	10

Management's job is planning, organizing, directing, and evaluating

Neutral

Management's job is prediction through understanding the capability of processes and people as well as the interaction between them

J. I believe that:

1	2	3	4	5	6	7	8	9	10

Managers should not necessarily be leaders

Neutral

Managers should be leaders

A leader who manages in accordance with Dr. Deming's theory of management will have a score of 10 on each of the above questions and a total score of 100 points. The closer a manager's score is to 1 for a given question, the less a manager leads in accordance with the System of Profound Knowledge. A Quality Management expert can use a manager's scores on the above questions to pinpoint the elements of the System of Profound Knowledge that a manager is having difficulty internalizing.

After identifying the specific areas in need of attention, the expert and individual manager meet to discuss why holding a particular belief is detrimental to transformation. In depth, private sessions may be necessary to understand the "whys" and "ramifications of" and "alternatives to" the above beliefs. The behaviors that these beliefs promote are also discussed, and ways of changing both attitudes and behaviors are suggested.

Identifying and resolving personal barriers is one of the most important areas in the transformation of an organization because it addresses the root cause of the most common reason for failure, lack of commitment on the part of top management. The following section presents an example of a private session between the Quality Management expert and Bob, the CEO of the Universal Co.

QM EXPERT:	I wanted to meet with you individually because I sensed an area that you're not entirely comfortable with.
BOB:	Let's get something out in the open. What is it?
QM EXPERT:	Promoting *cooperation* instead of competition.
BOB:	(*Thinks for a moment*) You're pretty sharp.
QM EXPERT:	That's what you pay me for. Anyway, it's an area that a lot of top guys have trouble with. After all, you got where you are by being the best you can be and beating out other people. It follows that you want to keep competing, and encourage your people to do the same.
BOB:	I guess.
QM EXPERT:	When you were in high school, did you play a sport?
BOB:	Yeah, football.
QM EXPERT:	Did you play varsity?
BOB:	No, and I was miserable about it. I really wanted that letter, but the coach wouldn't put me on the team.
QM EXPERT:	So you remember how it felt?
BOB:	Like it was yesterday. I hated it.
QM EXPERT:	What did you hate?
BOB:	I was trying as hard as the guys who made the team. And, I'm not so sure the guys on the team were better than I was.
QM EXPERT:	Exactly, that's Dr. Deming's point. Competition doesn't help anybody improve. It just creates winners and losers. Do you think you're creating this kind of situation anywhere for our people?
BOB:	I guess I am. We have an "Employee of the Month" contest. Only one person wins. I'm sure there are others who deserve to win, too. Their morale probably hits the floor when they don't win.

QM EXPERT: I'm sure you're right. Now that you understand the downside of competition at a gut level, I know we'll be able to work on improvement in this area.

BOB: Thanks for the feedback. I don't think I would have come to this by myself.

QM EXPERT: That's what I'm here for. I'm glad I could help.

THE QUALITY MANAGEMENT LEADER

Top management's education and self-improvement includes group and individual study of the System of Profound Knowledge and the identification and resolution of personal barriers to transformation. Guided by an expert, the Quality Management leader will develop his or her management style to incorporate the attributes necessary to lead the organization.* The Quality Management leader should possess the following characteristics:

1. A leader sees the organization as a system of interrelated components, each with an aim, but all focused collectively to support the aim of the system of interdependent stakeholders. This type of focus may require suboptimization of some components of the system.
2. A leader tries to create for everybody interest and challenge, joy in work, and pride in the outcome. He or she tries to optimize the education, skills, and abilities of everyone, and helps everyone to improve. Improvement and innovation are his or her aim.**
3. A leader coaches and trains, and does not judge and punish.*** He or she creates security, trust, freedom, and innovation. A leader is aware that creation of trust requires that he or she take a risk.**** A leader is an active listener and does not pass judgment on those to whom he or she listens.
4. A leader has formal power, power from knowledge, and power from personality. A leader develops and utilizes the power from knowledge and personality when operating in an existing paradigm of management. However, a leader may have to resort to the use

---

* The attributes of a leader are paraphrased from Deming, W. E., *The New Economics*, 2nd ed., 1994, pp. 125–128. The author takes sole responsibility for any errors introduced in restating Dr. Deming's words.
** Deming, *New Economics*, p. 125.
***Ibid., p. 126.
****Carlisle, and Parker, *Beyond Negotiation*, John Wiley & Sons (New York), 1989.

of formal power when shifting from one paradigm of management to another.

5. A leader uses plots of points and statistical calculation with knowledge of variation, to try to understand both the leader's performance and that of his or her people. A leader is someone who knows when his or her people are experiencing problems that make their performance fall outside of the system and treats the problems as special causes of variation. These problems could be common-cause to the individual (long-term alcoholic) but special-cause to the system (alcoholic works differently from his peers).

6. A leader understands the benefits of cooperation and the losses from competition.*

7. He or she does not expect perfection.

8. A leader understands that experience without theory does not facilitate prediction of future events. For example, a leader cannot predict how a person will do in a new job based solely on experience in the old job. A leader has a theory to predict how an individual will perform in a new job.

9. A leader is able to predict the future to plan the actions necessary to pursue the organization's aim. Rational prediction of future events requires that the leader continuously work to create stable processes with low variation.

A manager who does not possess all of the above attributes will have problems with the transformation. Again, it may be necessary to arrange for private sessions between some members of the EC and an expert in Dr. Deming's theory of management to discuss in depth, and confidentially, the "whys" and "ramifications" of the above attributes. As with beliefs, this step may be critical to a successful transformation because it goes to the root cause of the most common reason for failing to transform an organization, lack of commitment on the part of top management.

DECISION POINT

Once the "neck" phase is well under way, the members of the EC face their last critical decision point in the fork model. If the members of the EC discover that they cannot overcome their difficulties with Deming-based Quality Management, then a "*NO GO*" decision is made and all

---

* Deming, W. E., *New Economics*, p. 128 and Kohn, A., *No Contest: The Case against Competition*, Houghton Mifflin (Boston), 1986.

efforts toward Deming-style Quality Management stop. On the other hand, if the members of the EC discover that they can overcome their difficulties with Deming-based Quality Management, then a *"GO"* decision is made and the Quality Management effort proceeds to the prongs of the fork model. These prongs begin with

1. Selecting initial process improvement leaders in the departments (see step 14 of the Detailed Fork Model in Figure 1.2).
2. Selecting initial cross-functional process improvement projects to address issues concerning transformation that span departments within the organization (see step 21 of the Detailed Fork Model in Figure 1.2).
3. Conducting an initial presidential review of the policy of the organization (see step 28 of the Detailed Fork Model in Figure 1.2).

SUMMARY

Chapter 3 presents a discussion of the Neck of the fork model that is presented in this book. The Neck is management's education and self-improvement. After top management of an organization commits to transformation, its members enter a period of education and self-improvement. These are difficult, soul-searching activities that have a profound effect on the individual and the organization.

One of the first tasks of the EC is forming one or more education, training, and self-improvement groups that concentrate on the following areas:

1. Studying the System of Profound Knowledge
2. Answering prescribed questions on quality management
3. Designating study teams for each of the 14 points
4. Identifying and resolving personal barriers to transformation

The System of Profound Knowledge is discussed by the Executive Committee under the guidance of a Quality Management expert. The expert creates an environment in which group members deepen their understanding of how the System of Profound Knowledge might affect organizational and personal decision-making. Chapter 3 provides an example of a group session, in dialogue form, to illustrate the role of the QM expert and the desired atmosphere of the group setting. Chapter 3 also presents several case studies and workshops that can be used by group members to deepen their understanding of the System of Profound Knowledge.

The second area in which a group can focus is answering, in writing, an appropriate selection of the 66 questions in Chapter 5 of Dr. Deming's book, *Out of the Crisis*. Group members reach consensus on their answers, and the answers are discussed with a QM expert. An example of this type of correspondence between a group and a QM expert is presented in Chapter 3.

Designating study teams for each of the 14 points is another task of the EC. The job of each study team is to remove organizational barriers to the transformation by studying and using the System of Profound Knowledge. An example is a study team that is working on point 12, "Remove barriers that rob employees of their pride of workmanship." The cross-functional team identifies the company's performance appraisal system as a barrier and uses the System of Profound Knowledge as the basis for discussing and redesigning the process.

The fourth area that is important in management's education and self-improvement is identifying and resolving personal barriers to transformation. Each member of the EC examines his or her opinions to determine if he or she holds any beliefs that would be detrimental to the transformation. The QM expert can use a questionnaire to help pinpoint problem areas that individuals may be experiencing.

Identifying and resolving personal barriers is one of the most important areas in the transformation, because it addresses the root cause of the most common reason for failure, lack of commitment on the part of top management. In-depth, private sessions may be necessary between the QM expert and an individual manager. An example of such a session is provided in Chapter 3.

Characteristics of the Quality Management leaders are discussed. Certain attributes are necessary if the leader hopes to transform an organization. Guided by an expert, the leader will develop these traits through the use of the methods presented in this chapter.

4

PRONG ONE:
DAILY MANAGEMENT

PURPOSE OF THIS CHAPTER

The purpose of this chapter is to explain what is required to develop, standardize, deploy, maintain, improve, and innovate the methods required for daily work in all areas of an organization. Daily work is managed through "daily management," Prong One of the model presented in this book.

When top management is ready to begin transforming the organization, it needs concrete ways of translating theory into practice. Daily management is one of the vehicles top management uses to accomplish this task.

SELECTING INITIAL PROJECT TEAMS

The members of the EC select initial Process Improvement Leaders (PILs) in the different departments of an organization (see step 14 of the Detailed Fork Model in Figure 1.2). PILs can be either full-time or part-time. The decision to have only full-time PILs, only part-time PILs, or both types of PILs depends on the needs of project teams.

Early in a transformation, an organization may need a greater proportion of full-time PILs. However, as the transformation proceeds, an organization may need a smaller proportion of full-time PILs. The change in the need for full-time PILs is due to an increase in the general level of "quality" knowledge in the organization and, consequently, a decrease in the need for the aid provided to project teams by full-time PILs.

PILs receive training in "Basic Quality Control Tools"* and "Psychology of the Individual and Team"** (see step 15 of the Detailed Fork Model in Figure 1.2).

Next, the members of the EC select the initial projects to be addressed by project teams (see step 16 of the Detailed Fork Model in Figure 1.2). Once the PILs and the projects have been determined, the members of the EC, in consultation with the PILs, select the members of each initial project team (see step 17 of the Detailed Fork Model in Figure 1.2). Project teams are formed with a specific purpose, consist of people from the same area (small unit), and exist in perpetuity. All project team members receive training in "Basic Quality Control Tools"*** and "Psychology of the Individual and Team"**** (see step 17 of the Detailed Fork Model in Figure 1.2).

DOING DAILY MANAGEMENT

Overview

After training, each initial project team works on one or more methods through daily management (see step 18 of the Detailed Fork Model in Figure 1.2). Daily management is developing, standardizing, deploying, maintaining, improving, and innovating the methods required for daily work. The development, standardization, and deployment of methods for daily work is called *housekeeping,***** because "it is a procedure which sets things in order."****** The maintenance, improvement, and innovation of methods for daily work is simply called *daily management*. Once again, the reader is cautioned that "daily management" is used in two different contexts in this section. First, it describes developing, standardizing, deploying, maintaining, improving, and innovating the methods required for daily work. Second, it describes only the maintenance, improvement, and innovation of methods for daily work.

---

* An excellent text for this seminar is Gitlow, H., Oppenheim, A., and Oppenheim, R., *Quality Management: Tools and Methods for Improvement*, 2nd ed., Irwin (Burr Ridge, IL), 1995.

** An excellent text for this seminar is Scholtes, P., *The Team Handbook: How to Use Teams to Improve Quality*, Joiner Associates (Madison, WI), 1988.

***Excerpts from Gitlow, Oppenheim, and Oppenheim, Quality Management are appropriate for this seminar.

****Scholtes, *The Team Handbook* is appropriate for this seminar.

***** The name "housekeeping" is taken from Total Productivity Maintenance (TPM). See Imai, M., *KAIZEN: The Key to Japan's Competitive Success*, Random House (New York), 1986, pp. 158–159.

******Schultz, L., *Process Management International* (Minneapolis, MN), 1990.

Housekeeping

The housekeeping functions of daily management are developed through a procedure called *function deployment.** Function deployment requires that relevant employees determine what functions are required to perform each method needed in their daily work. Each function is subject to the scrutiny of the following questions:

1. Why is this function required?
2. What is this function intended to achieve? What is the aim of this function?
3. What resources are necessary for this function?
4. What target must be set to allocate appropriate resources to this function to optimize the aim of the organization?
5. Where in the process should this function take place?
6. When should this function be implemented?
7. Who is responsible for this function?
8. How does this function contribute to the optimization of the system of interdependent stakeholders for the organization?
9. What measurements are used to monitor this function?
10. How will this function be carried out?
11. Does this function contain non-value-added steps?

Housekeeping is practiced through the SDSA cycle. As we discussed earlier, the SDSA cycle has four stages: "Standardize," "Do," "Study," and "Act." The Standardize stage involves teaching employees how to study and understand the causal factors that affect each critical method with which they work (e.g., using flowcharts). The employees developing the best practice method use the flowchart to highlight non-value-added steps, and work toward eliminating them. Employees can also use other tools to understand the causal systems that affect their methods, such as cause and effect diagrams, interrelationship diagrams, and simulations. All the employees who use a particular method compare notes on causal factors and develop one "best practice method," as seen through a "best practice" flowchart.

At this stage, question 11 above can be addressed by using an integrated flowchart. This type of flowchart adds at least one dimension to a typical flowchart. An example of an integrated flowchart with one extra

* Mizuno, S., *Company-Wide Quality Control*, Asian Productivity Organization, 4-14, Asasaka 8-chome, Minato-ku, Tokyo 107, Japan, 1988, pp. 55–61.

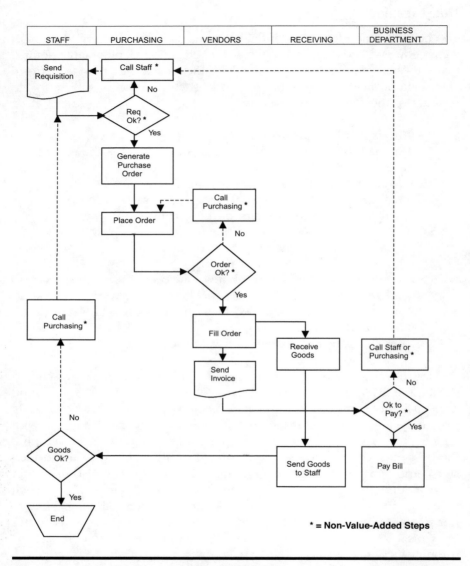

STAFF	PURCHASING	VENDORS	RECEIVING	BUSINESS DEPARTMENT

Figure 4.1 Integrated Flowchart Showing Value Added and Non-Value-Added Steps

dimension is shown in Figure 4.1. The non-value-added steps are indicated with an asterisk.

An integral part of preparing a "best practice" method is developing key indicators to monitor the "best practice" method. Key indicators can yield data that is measurable or nonmeasurable. Nonmeasurable data, also called "unknown and unknowable" data,* frequently include the most important business figures, such as the cost of an unhappy customer or the benefits of a prideful employee. It is not accurate to assume that if a process cannot be measured, it cannot be managed. Nonmeasurable data, like interactions with other people, are managed on an ongoing basis.

The Do stage entails a project team conducting a planned experiment to collect measurements on key indicators for determining the optimal configuration of the "best practice method" on a trial basis. The Study stage consists of project team members studying the measurements on the key indicators for determining the effectiveness of the "best practice method." The Act stage involves the establishment of a standardized best practice method, using a best practice flowchart. This is then formalized by training all relevant employees in the best practice method and by updating training manuals.

A best practice method can be quite complex. It can be constructed to take into account a great number of contingencies. For example, if a customer has complaint A, follow method A; however, if a customer has complaint B, follow method B, and so on. Or, if a customer has complaint A and claims it is urgent, follow method A1; however, if a customer has complaint A and places no urgency on the matter, follow method A2.

Measurements on Key Indicators

Best practice methods are monitored through measurements taken on key indicators. Key indicators possess two important characteristics that make them useful in a system of quality management. First, they are operationally defined. They have a uniformly agreed-upon definition,** which promotes communication between people. Second, they monitor "results" and the "processes that generate results."

Key indicators are either results-oriented or process-oriented. Results-oriented key indicators, called *R criteria*, are used to evaluate the results of a method. They are called *control points* (also "check points"). Process-oriented key indicators, called *P criteria*, are used to evaluate a method that creates results. They are called *control items* (also "check items"). As one author put it, "P criteria call for a long-term outlook, since they are directed at people's efforts and often require behavioral change. On the

---

* The notion of "unknown and unknowable" control items is taken from Deming, W. E., *Out of the Crisis*, M.I.T. Center for Advanced Engineering Studies (Cambridge, MA), 1986, pp. 121–122.
** Gitlow, Oppenheim, and Oppenheim, *Quality Management*, pp. 64–67.

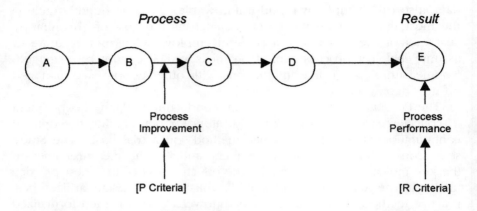

Figure 4.2 Relationship between P and R Criteria

other hand, R criteria are more direct and short term."* Figure 4.2 depicts
the relationship between P criteria and R criteria.

The relationship between control points and control items is shown
in the following example. Adherence to safety policies is a control item
(P criteria), while the number of injuries per 100 employees per month
is a control point (R criteria).

The relationship between manager and subordinate can be defined in
terms of control items and control points. A control item for a manager (e.g.,
safety) is measured or evaluated at a control point by a subordinate (e.g.,
number of injuries per 100 employees per month). In this way, an interlocking
set of R and P key indicators is developed throughout an organization.

Daily Management

After a best practice method has been developed and deployed by a
project team, housekeeping activities give way to daily management
activities. Daily management is used to determine the actions necessary
for a project team to maintain, improve, or innovate methods. Daily
management is performed to decrease the difference between process
(actual) performance and customer requirements. A process with a large
variance not centered on a desired customer requirement creates a prob-
lematic difference between process performance and customer needs.
Daily management is needed to reduce process variation and center the
process on the nominal (the desired customer requirement).

* Imai, M., *KAIZEN*, p. 18.

Daily management is accomplished by using the PDSA cycle. As discussed earlier, the PDSA cycle* consists of four stages: "Plan," "Do," "Study," and "Act." A Plan is developed to improve or innovate a standardized "best practice" method (Plan). A plan can take the form of a modified "best practice" flowchart.

Ideas for improvement or innovation of a process (as conceived of as a flowchart) come from study of the causal factors that affect it. There are many tools that can be used to help employees understand causal factors. The tools fall into two groups. The first group includes data-based decision-making tools such as check sheets, Pareto diagrams, histograms, run charts, control charts, and scatter diagrams, to name a few. The second group includes proven change concepts** such as incorporating technology into the process, shifting demand patterns to off-peak times in a process, reducing controls in a process, performing tasks in parallel in a process, conducting training in a process, outsourcing steps of a process, and benchmaking. Both sets of tools assist an individual or team doing daily management to modify the existing best practice flowchart to a revised best practice flowchart.

The Plan is monitored by taking measurements on key indicators on a small-scale or trial basis, and tested through a planned experiment by project team members (Do). The effects of the Plan are studied (Study) by examining measurements on key indicators, and appropriate corrective actions are taken (Act). These corrective actions can lead to a new or modified Plan, or are formalized through training all relevant employees and updating training manuals. The PDSA cycle continues forever in an uphill progression of never-ending improvement.

It is important to note that the success of another person or organization is not a rational basis for turning the PDSA cycle. For example, isolating one component of System A and expecting it to work within the context of System B is not rational. The reasons behind the success in System A may not be present in System B. Therefore, copying without a true understanding of the conditions (causal factors) surrounding the copied system can lead to misapplication of the PDSA cycle. For example, an electric utility copying a customer service process from a manufacturing company, without understanding the reasons for the success of the customer service process in the manufacturing company, can lead to a poorly conceived revised best practice method for the electric utility, and hence, a misapplication of the PDSA cycle.

---

* Deming, W. E., *Out of the Crisis*, pp. 86–89.
** See Langley, G., Nolan, L., Nolan, T., Norman, C., and Provost, L., *The Improvement Guide: A Practical Approach to Enhancing Organizational Performance*, Jossey-Bass (San Francisco), 1996, pp. 295–359.

Daily Life Example of Daily Management

Bart's exercise regimen is important to him. He realizes that he is not exercising as much as he would like. He collects data on his exercise habits for a period of 8 weeks. The data from his initial investigation are shown in Figure 4.3.

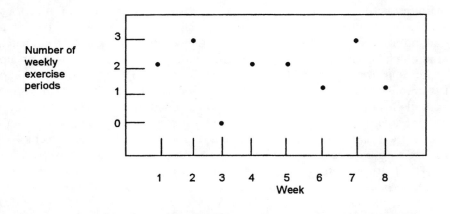

Figure 4.3 Initial Analysis of Exercise Habits

Analysis of the data leads Bart to question his method for "making exercise happen." He realizes that he has no method, so he develops the flowchart shown in Figure 4.4.

Bart discusses his exercise method with his physician during week 9 of his exercise process. His physician states, based on medical knowledge, that Bart should exercise for 40 minutes, at least three times per week. Thus, he establishes a target of three exercise periods per week. The measurement for this method is the number of 40-minute exercise periods per week. Bart records the number of 40-minute exercise period per week. The record is shown in Figure 4.5.

The record shows that the target of three exercise periods per week is achieved for weeks 10 through 15, but not in weeks 16 and 17. This leads Bart to go back and examine his method. In so doing, he discovers that the reason he failed to exercise three times in weeks 16 and 17 was that he had no notation for exercise in his appointment book in weeks 16 and 17. He realizes that his exercise method has to be changed to prevent this problem from happening in the future. He revises his method to add in a notation to "write in more exercise periods" after his last noted exercise period. This revision to his method is shown in Figure 4.6.

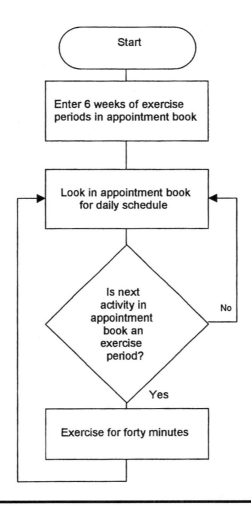

Figure 4.4 Flowchart of Exercise Program

Bart collected more data and consistently met his target for weeks 18 through 28; see Figure 4.7.

Business Example of Daily Management*

The following business example may use some statistical tools with which you are not familiar. Don't worry about it. Just read the case study to

---

* Krishnan, K. and Gitlow, H., "Quality Improvement in the Treatment of Cold Gas Plasma: A Case Study," *Quality Engineering*, 9(4), 1997, pp. 603–614.

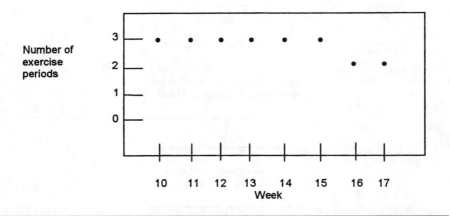

Figure 4.5 Analysis of Exercise Habits

get a "feel" for how daily management and the PDSA cycle work in a business setting.

Cold gas plasma treatment consists of exciting gas molecules to strip and recombine the electrons in the surface of a polymer. By varying the conditions of cold gas plasma treatment, it is possible to obtain a particular effect on the surface of a polymer, such as superior bonding, printing, potting, or wetting.

In this study, cold gas plasma treatment is used to improve the wettability of the surface of cuvettes. Cuvettes are molded plastic containers used to hold a sample for analysis. Wettability allows for accurate analysis of the sample. Poor wettability causes material from sample i to remain in the cuvette after the introduction of sample $i + j$, where $j = 1$ to m. This can result in "carryover" error.

Operational Definition of Wettability

A surface is wettable when a liquid has the ability to spread on it. Wettability is measured by meniscus size. A meniscus can be concave, convex, or flat; see Figure 4.8. Superior wettability is exhibited by a large concave meniscus.

Measurement of a meniscus is shown in Figure 4.9. The distance X is obtained by squirting distilled water into a cuvette that is held upright at eye level such that point A is between 0.25 and 0.50 in from the base of the cuvette. X is measured by using an optical eyepiece (Mitutoyo 183 or equivalent) with a millimeter scale with gradations of 0.10 mm and distilled water at room temperature (72°F ± 10°F).

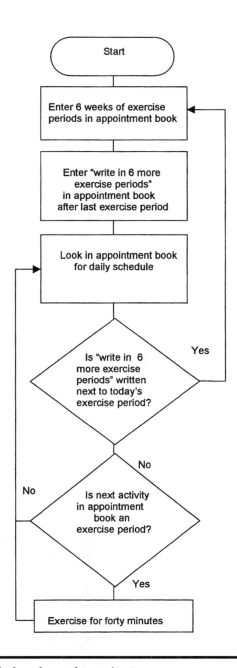

Figure 4.6 Revised Flowchart of Exercise Program

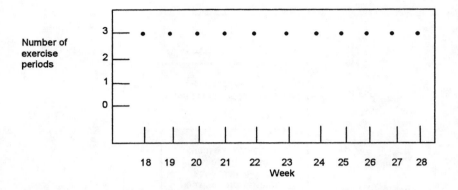

Figure 4.7 Continuing Record of Exercise

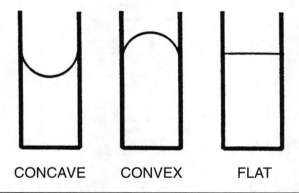

CONCAVE   CONVEX   FLAT

Figure 4.8 Illustration of Meniscus

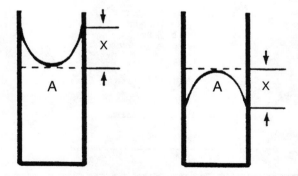

Figure 4.9 Measurement of Meniscus

The specification for an acceptable meniscus is 2.4 mm to 3.2 mm; desired level (nominal) is 2.4 mm, acceptable lower deviation from nominal (lower tolerance) is 0.0 mm, and acceptable upper deviation from nominal (upper tolerance) is 0.8 mm. Meniscus below 2.4 is not acceptable. Meniscus over 3.2 is not acceptable, because it increases cost from unnecessary plasma treatment and degrades the surface of the cuvette due to excessive molecular activity.

Background Information

The cold gas plasma treatment method has been employed in the process under study for over 5 years. Discussions with the process engineer, area supervisor, and operators revealed that the cold gas plasma treatment process produced output that met specifications most of the time. If the meniscus size was below the lower specification limit, the cuvette was rerun. If the meniscus size was over the upper specification limit, the cuvette was accepted. Operators measured, but did not record, meniscus measurements. Operators collected data for a control chart analysis of the cold gas plasma treatment process. The operators filled in a process control data sheet (Figure 4.10) and recorded meniscus measurements directly onto a control chart.

Date: _____   Time in: _____   Time out: _____

Part Number: 6603624   Lot Number: _____

Lid number: _____   Bottom number: _____

(In) Vacuum: _____   (Out) Vacuum: _____

Quantity: 70   Quantity inspected: 5   Inspected by: _____

Figure 4.10 Process Control Data Sheet; Cuvette

Process Flowchart

A flowchart of the cold gas plasma treatment process prior to control chart analysis is shown in Figure 4.11. In most cases, a cuvette would be reprocessed if its meniscus specification was below the lower specification limit. This is reflected by the loop in the flowchart. Usually, the lower specification limit would be met with two or three loops through the process. If the equipment was malfunctioning, the process engineer was called to take corrective action.

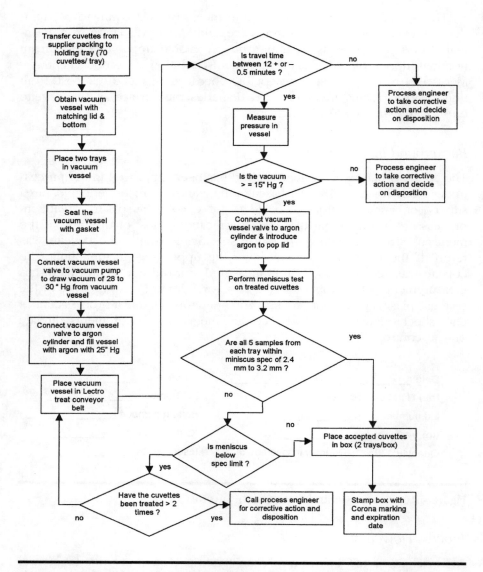

Figure 4.11 Flowchart of Original Cold Gas Plasma Treatment Process

Turns of the Plan-Do-Study-Act Cycle

The Plan-Do-Study-Act (PDSA) cycle was turned twice in this case study. The first turn stabilized the cold gas plasma treatment process. The second cycle improved the capability of the cold gas plasma treatment process.

PDSA Cycle One. The Plan phase consisted of understanding and ana-lyzing the present situation by making the flowchart of the current process (see Figure 4.11) and collecting meniscus measurements from 20 subgroups of cold gas plasma treated cuvettes for an $\bar{x}$ and R chart (see Figure 4.12). Each subgroup consisted of a random sample of 5 cuvettes from a tray of 70 cuvettes. Each production run consisted of 20 trays. The cuvettes within a tray are assumed to be homogeneous by the process engineer.

The R chart indicated a stable process with respect to variation. There were no out-of-control points. The $\bar{x}$ chart showed six out-of-control points. The out-of-control points were analyzed by the process engineer and the operators. They developed a cause-and-effect diagram to identify the possible factors that could cause out-of-specification meniscus; see Figure 4.13. The cause and effect diagram was used to study the six out-of-control points on the $\bar{x}$ chart. The process engineer and operators could not assign any special causes to out-of-control points 1–4. However, it was quickly realized that a power interruption caused out-of-control points 5 and 6. These subgroups were from one vacuum vessel that was inside the cold gas plasma treatment equipment during the occurrence of a power interruption. As this was a special cause, a new policy was instituted concerning electrical failure of the cold gas plasma treatment equipment; see Figure 4.14.

The operator was asked why the two trays indicated by out-of-control points 5 and 6 were not identified in the comments section of the log as having occurred during a power interruption. The operator said that due to her limited English-speaking abilities, she did not know what to write in the log. A new policy that log sheet comments could be written in English or Spanish was established by the area supervisor.

Assuming the new policies were in place, out-of-control points 5 and 6 were dropped and the $\bar{x}$ chart was recalculated; see Figure 4.12. Now, the $\bar{x}$ chart indicated a stable process. Please note that out-of-control points 1–4 were now in control. These points only looked out of control because of out-of-control points 5 and 6.

The Do phase consisted of testing the two new policy guidelines during full-scale operation of the cold gas plasma treatment process.

The Study phase consisted of determining if all the operators knew what to do in case of a power interruption and if all the operators wrote their process comments on the log sheet in English or Spanish. The effectiveness of the power interruption policy could not be verified, because there were no further power failures during the course of this study.

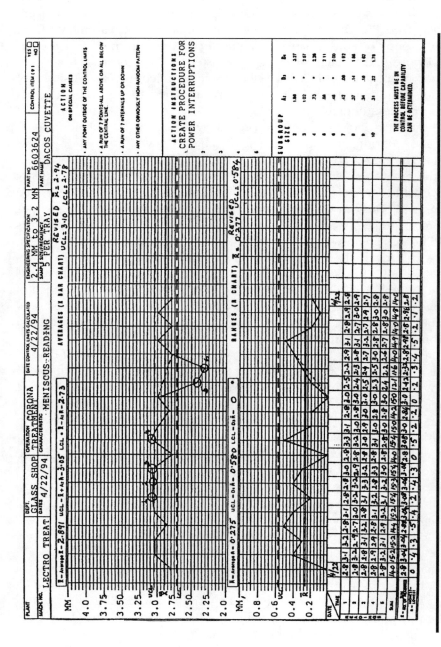

Figure 4.12 $\bar{x}$ and R chart — April 23, 1994

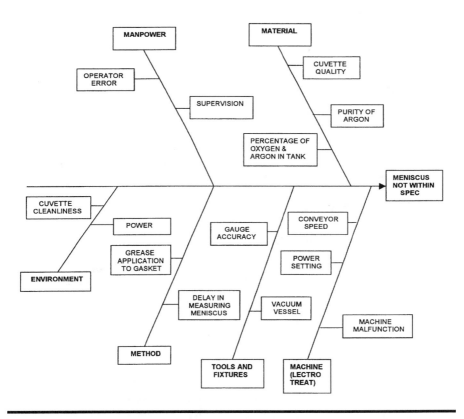

Figure 4.13 Cause-and-Effect Diagram; Reasons for Meniscus to Fall Outside of Specific Limits

The Act phase consisted of updating the procedure manual to ensure that the new policies are followed by operators; see Figure 4.14. The next time the process was run, it was in control except for one point that was just below the lower control limit on the $\bar{x}$ chart; see Figure 4.15. The process engineer and operators could not determine the cause of this out-of-control point.

PDSA Cycle Two. The Plan phase consisted of monitoring the process by plotting 20 new subgroups on the existing $\bar{x}$ and R charts; see Figure 4.16. Note that the control limits are dashed lines, indicating that they are projected from the prior $\bar{x}$ and R charts; Again, the R chart indicates a stable process. However, a cyclical pattern was observed on the $\bar{x}$ chart. It was indicated by every fifth and sixth subgroup being below the lower

Corona Treatment of Cuvettes

7.16 Close cuvette vessel valve.
7.17 Disconnect argon/oxygen vacuum from cuvette vessel valve.
  7.17.1 Record "In Vacuum" on the *Process Control Data Sheet*, Figure 3.
  7.17.2 Record "In Vacuum" on *Cuvette Vessel Usage Log*, Figure 2.
7.18 Immediately place vessel on Lectro-Treat conveyor belt.

  Note: Vessels must be placed onto the center of the conveyor in single file one behind the other, and with the valve end of the vessel facing the conveyor entrance. Spacing is not a concern.

  Note: Vessels *must not* be placed side by side.
  7.18.1 Log treatment time in and time out on Process Control Data Sheet.
7.19 Travel time for a vessel must be 12 ± .05 min.
7.20 Remove vessel from Lectro-Treat.
  7.20.1 In the event of electrical failure during the vessel treatment process, the cuvettes must be removed and scrapped, MRR disposition is to be performed on a weekly basis.
7.21 *Immediately* attach the 30 in. Hg vacuum gauge to vessel.
  7.21.1 Open vessel valve and note reading. If vacuum is equal or greater than 15 in. Hg, then record "Vacuum Out" on *Process Control Data Sheet.*

Figure 4.14 Procedure Manual Reflecting Policy Changes for Electrical Failure

control limit. In the past, the operators would not have noticed anything abnormal because the individual meniscus measurements were within specification limits.

An analysis was done by the process engineer using the previously developed cause-and-effect diagram (see Figure 4.13). In the process engineer's opinion, the cyclical pattern on the $\bar{x}$ chart indicated that the meniscus problem was related to one of the vacuum vessels used in the cold gas plasma treatment process. Consequently, the process engineer expanded the subcauses for "vacuum vessel" on a new cause-and-effect diagram; see Figure 4.17.

Vacuum vessel 021 was identified from the log sheets as the troublesome vacuum vessel. The process engineer concluded that the cause of the problem was a leak either in the gasket area or in the valve area of vacuum vessel 021.

The Do phase consisted of testing vacuum vessel 021 for leaks in the gasket and valve areas. It was determined that the leak was in the

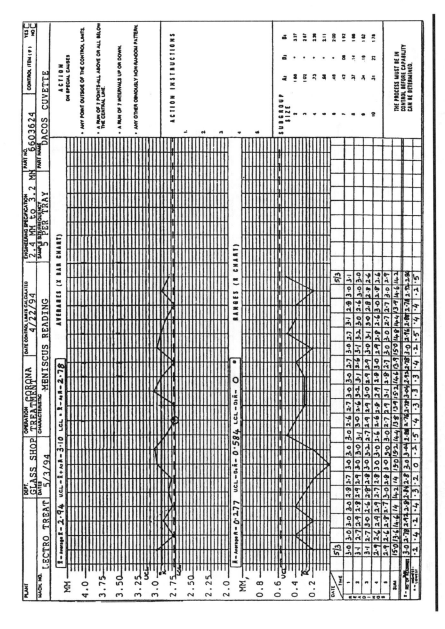

Figure 4.15 Recalculated $\bar{x}$ and R chart — May 3, 1994

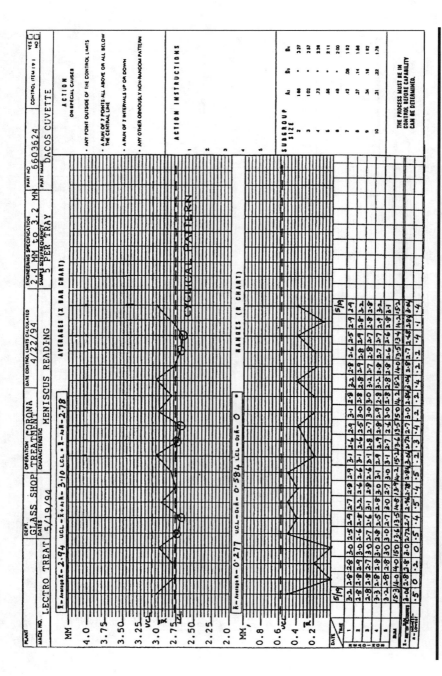

Figure 4.16 Projected x̄ and *R* chart — May 19, 1994

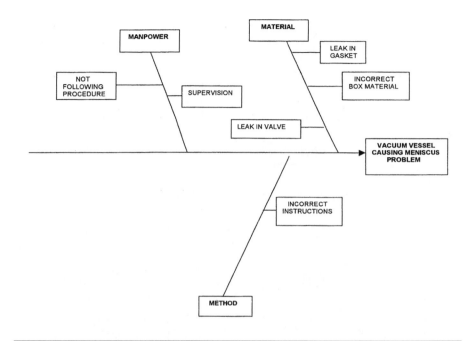

Figure 4.17 Cause-and-Effect Diagram; Reasons for Meniscus Problems Due to Vacuum Vessel

gasket seating area of the vessel. Vacuum vessel 021 was scrapped in conformance with company procedure and a new vessel was installed in its place.

The Study phase consisted of sampling 24 additional subgroups to determine the stability of the cold gas plasma treatment process. New $\bar{x}$ and R charts indicated that it was stable (Figure 4.18).

The Act phase consisted of changing the procedure manual for the cold gas plasma treatment process to ensure that there is a standardized method for dealing with leakage in vacuum vessels; see Figure 4.19. The revised procedure ensures that the vacuum is maintained for 2 min in the vacuum vessel. If a vacuum cannot be maintained for a full 2 min, the procedure manual states the appropriate course of action to be taken by the operator.

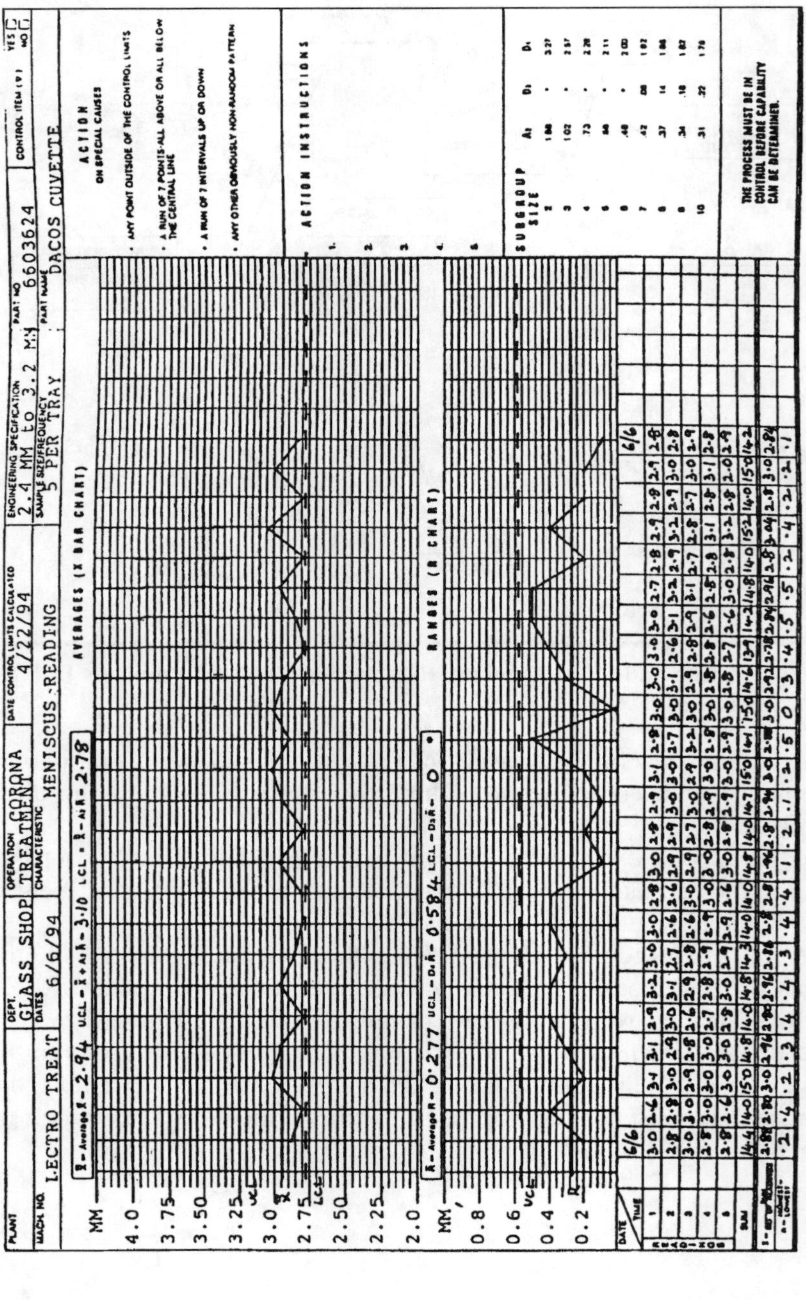

Figure 4.18 $\bar{x}$ and R chart — June 6, 1994

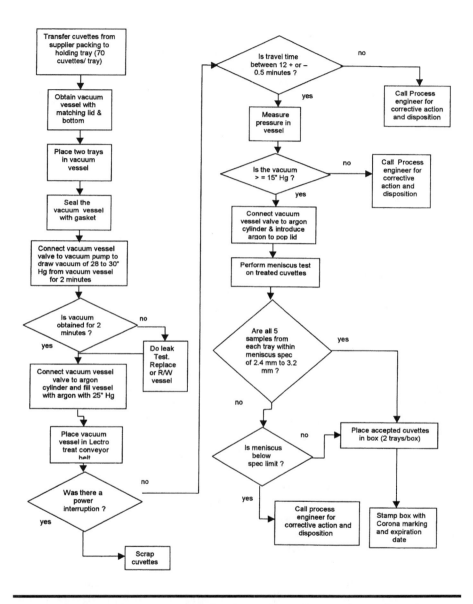

Figure 4.19 Procedure Manual Reflecting Policy Changes for Leakage in Vacuum Vessel

Improved Process

The two turns of the PDSA cycles resulted in a revised method for cold gas plasma treatment; see Figure 4.20. Based on the last 24 data points, the revised method is stable; its capability is indicated by Z_{USL} = 2.47 and Z_{LSL} = 4.02, and its upper natural limit (UNL) is 3.264 mm and lower natural limit (LNL) is 2.526 mm, compared to the specification values of 3.2 and 2.4, respectively.

Conclusion

This study produced several benefits. First, it produced benefits to the internal customers of the cold gas plasma treatment process in the form of reduced network costs from not recycling cuvettes and decreased surface degradation to cuvettes due to fewer cold gas plasma treatments. Second, it yielded benefits to the external customers of the final product (a chemical instrument) through increased on-time delivery, decreased scrap rates, and increased quality. In view of the improvements created in this case study and the simplicity of maintaining a hand-drawn control chart, the area supervisor and the process engineer will continue the use of the control chart.

Management Review

Project teams present their housekeeping and daily management projects to managers for approval in management reviews. A management review* involves comparing the actual results generated by applying a set of methods with the targets established to allocate resources to optimize the organization toward its aim, and finding opportunities to improve and innovate methods.

Three critical inputs are required for a management review. They are a well-researched method (called a "best" practice method), a target established to allocate resources to optimize the aim of the organization, and an actual result that has been measured through a control point. The development of the first and second inputs requires that a manager have a deep and thorough understanding of the method being studied, a firm grasp on where the method stands in respect to process capability and environment, knowledge of the aim of the organization to determine appropriate methods to get there,** realization that a method is used to predict a result, recognition that a method should yield a high likelihood

---

* Mizuno, *Company-Wide Quality Control*, pp. 269–280.
** Ibid., p. 98.

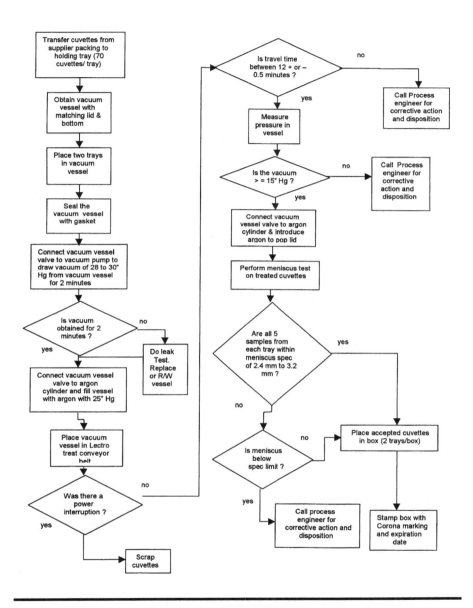

Figure 4.20 Flowchart of Revised Cold Gas Plasma Treatment Process

of achieving a target before it is implemented,* and understanding that targets are vehicles for allocating resources between methods.

A set of suggested questions that can form the basis of a management review of a project team is listed below. These questions will help all stakeholders involved in a management review focus on opportunities for improvement and innovation of methods. These questions are only suggestions. Management reviews have natural flows. A manager can use preset questions, but also needs to go with the rhythm of the review if he or she is to accomplish the purpose of the review.

1. What is your group's first priority method?
2. Are you (as an individual) or your group working on improvement or innovation of your first priority method?
3. How do you measure the performance of the first priority method? What are the control items and control points for this method?
4. What is your group's target for this method? Monthly? Yearly?
5. Did you study this method last year? How have you incorporated the results of that study into your current method?
6. What is the status of your group's method to date?
7. Are methods yielding targets? Monthly? Yearly?
8. If targets are not being achieved, what countermeasures have your team members taken, and what actions will prevent the same situation from recurring in the future?

A management review probes the root cause(s) of the differences between actual results and targets without "tampering" with methods.

A management review includes a questioning process that asks questions "one inch wide and one mile deep," as opposed to questions that are "one mile wide and one inch deep." This means that the management review probes root causes to a high level of detail. A technique that helps people probe for root causes in the above manner is the "5W1H" process.** The "5W1H" process is used to ask "why" a problem occurs five times and then "how" the problem can be resolved, as opposed to just asking "how" the problem can be resolved. Historically, in the Western world, a person asks a question such as, "Why didn't the lawn get mowed this week?" and gets an answer such as, "The mower broke." This usually leads to the person responsible for mowing the lawn being blamed and no improvement of the lawn-mowing process. The "5W1H" process suggests something like the following:

---

* Adapted from Mizuno, p. 99.
** Imai, M., *KAIZEN*, p. 235.

Sample "5W1H" Process

Question 1:	"*Why* didn't the lawn get mowed this week?"
Answer 1:	"The mower broke."
Question 2:	"*Why* did the mower break?"
Answer 2:	"The bearing burned out."
Question 3:	"*Why* did the bearing burn out?"
Answer 3:	"The bearing burned out because it wasn't oiled properly."
Question 4:	"*Why* wasn't the bearing oiled properly?"
Answer 4:	"The bearing wasn't oiled properly because the oil line was clogged."
Question 5:	"*Why* was the oil line clogged?"
Answer 5:	"The oil like was clogged because there is no routine and proactive maintenance program to examine the oil line."
Question 6:	"*How* can we resolve this problem so it doesn't happen again?"
Answer 6:	"Develop and follow a policy of routine and proactive maintenance for the oil line."

As you can see, questions 1 through 5 focus on the root cause ("Why") of the problem, while the last question focuses on "How" to improve a process. The above example is an application of the "5W1H" process. The "5" and "1" are just symbolic numbers, which promote asking questions that are "one inch wide and one mile deep."

Variance Analysis

It is critical that management reviews be conducted in accordance with Dr. Deming's theory of management. All sources of variation are not due to special causes. A manager following Dr. Deming's theory of management doesn't tamper with the processes under his or her control. Instead, causes of variation are separated into common and special sources by statistical methods. Then, employees work to resolve special sources of variation, and management works to remove common sources of variation by modifying methods.

The management review focuses on whether the actual method (the method actually used by an employee) followed the best practice method. Figure 4.21* shows the relationship between following methods and achieving targets. Cell 1 shows the outcome of an employee following a

* The matrix was discussed by Dr. Noriaki Kano on April 1, 1990 in Atlanta, GA.

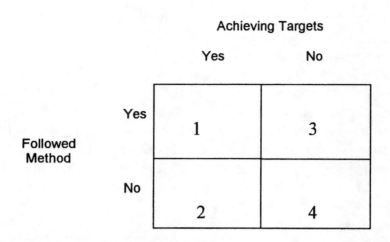

Figure 4.21 Relationship between Following Methods and Achieving Targets

best practice method as the attainment of a target. Cell 4 shows the outcome of an employee not following a best practice method as the failure to attain a target. To reverse this failure, the employee follows the best practice method. In this case, the management review determines answers to the following questions:

1. What best practice method was not followed?
2. Who failed to follow the best practice method? Note: The focus is on system problems, not on the individual. This will promote joy in work and pride in the outcome.
3. Why didn't the employee follow the best practice method? Was it due to ignorance, misunderstanding, lack of training, negligence, problems with a machine, or problems with raw materials?
4. Should the best practice method be changed to resolve problems due to ignorance, misunderstanding, lack of training, negligence, problems with a machine, or problems with raw materials?

Cell 2 shows the outcome of an employee not following a best practice method as the attainment of a target. In this case, depending on prevailing pressures, the employee may adopt a slower pace when determining why the method used yielded the target.

Cell 3 shows the outcome of an employee following the best practice method as the failure to attain a target. In this case, the best practice method is improved or innovated, and/or a change is made in the target;

the employee is not blamed. This change is accomplished by asking the following questions.*

1. What best practice method missed its target?
2. How can the best practice method be changed to attain its target?
3. Must the best practice method be changed to resolve problems due to ignorance, misunderstanding, lack of training, problems with a machine, or problems with raw materials?
4. What target was missed?
5. How much was the target missed over time? Is the process under study stable? Will adjustment of the target result in tampering with the best practice method?
6. Why was the target missed? Was the target set incorrectly due to ignorance, lack of training, problems with a machine, problems with raw materials, management, or by guesswork?

Once these questions are answered, the necessary information may be available for improvement or innovation of the best practice method or change of the target. These questions focus on improvement and innovation of the best practice method, not on blame of the individual.

Frequently, it is not possible to investigate the negative scenarios presented in cells 2, 3, or 4 on a daily basis. One day may not provide enough time to perform all four stages of the PDSA cycle to achieve the desired improvement and/or innovation. A special procedure called the *Quality Improvement* (QI) *story* is needed to deal with pressing daily problems.

Quality Improvement Story

The Quality Improvement (QI) story** is an efficient format for presenting process improvement studies to management. QI stories standardize the reporting of process improvement efforts, help avoid logical errors in analysis, and make process improvement efforts easier to deploy company-wide. QI stories also promote management based on data and facts, as opposed to management based on guesswork and opinion.

The QI Story has seven steps (Figure 4.22). All sections of a QI story are clearly numbered and labeled so that they can be related to one of

* The information in this paragraph is adapted from comments made by Dr. Noriaki Kano, Science University of Tokyo, on April 1, 1990, in Atlanta, Georgia.
** Gitlow, H., Gitlow, S., Oppenheim, A., and Oppenheim, R., "Telling the Quality Story," *Quality Progress*, September 1990, pp. 41–46.

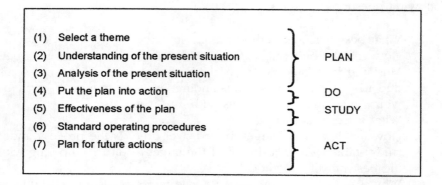

Figure 4.22 Relationship between the QI Story and the PDSA Cycle

the steps in a QI story. This seven-step procedure follows the PDSA cycle. The PLAN phase involves selecting a theme for the QI story based on all the background information necessary for understanding the selected theme (including a flowchart, the reason for selecting the theme, and organization, and department objective(s)); gaining a full understanding of the present situation surrounding the theme; and conducting an analysis of the present situation to construct a plan of action

The Do phase involves putting the appropriate plan into action on a small-scale or trial basis so that the process improvement actions can be tested through a designed experiment.

The Study phase involves studying, creatively thinking about, collecting, and analyzing data concerning the effectiveness of the plan experimentally set into motion in the Do phase. Does the plan reduce the difference between customer needs and process performance? Before-and-after comparisons of the effects of the experimental plan are presented in a QI story.

Finally, the Act phase involves determining if the plan was effective in reducing the difference between process performance and customer needs. If it was not effective, employees go back to the Plan stage to find other actions. If it was effective, employees either go to the Plan stage to seek the optimal settings of the actions or formally establish revised "best practice" methods (standard operating procedures) through training and updating training manuals. Further actions are taken to prevent backsliding for the plan set into motion. This step also includes identifying remaining process problems, establishing a plan for further actions, and reflecting on the positive and negative aspects of past actions.

Potential Difficulties

When using QI stories, difficulties may arise in two areas: qualitative (non-numerical) themes and exogenous problems. Qualitative themes are QI story themes that are difficult to describe with numerical values. For example, a qualitative theme for a QI story is "Improvement of Management Reviews." Qualitative themes are analyzed by focusing on the magnitude of the gap between actual performance and desired performance. Furthermore, they are improved based on decisions that stem from a manager's theory of management, not from data analysis. For example, the decision to improve the management review process is made on the basis of theory because no data exists on the negative effects of a suboptimal management review process.

Exogenous problems are problems that are seemingly beyond the control of anyone in the organization (e.g., electrical outages caused by power lines knocked out by lightning). It is important for employees to realize that it may be possible to take actions to remedy exogenous problems, as opposed to becoming overwhelmed and frustrated by them. When exogenous problems occur, employees analyze why there are so many occurrences of the exogenous problem in area A vs. area B, given both areas have equal opportunities for the occurrence of the exogenous problem. This may lead to the isolation of an action that decreases the difference between process performance and customer needs.

For years it was believed that electrical outages due to lightning striking power lines was an exogenous problem at Florida Power & Light Co. A QI story revealed that the incidence of outages in FP&L's South Dade district was much higher than in its West Broward district. Further study showed that the methods used for grounding power lines were different in both areas. This analysis led to the South Dade district's adopting the West Broward district's method for grounding power lines. Subsequently, the incidence of outages due to lightning dropped dramatically in the South Dade district. This seemingly exogenous problem was improved by the application of a QI story.

Pursuit of Objectives

Initially, QI stories will be selected because they are near-complete resolutions to department problems and will not relate to organization and department objectives. As employees gain experience with QI stories, they will want to select themes that do relate to organization and department objectives. If QI story activities are not consistent with department objectives, and department objectives are not consistent with organization objectives, there is the distinct possibility that quality improvement efforts will not be in line with organization objectives.

QI Story Case Study

A QI story drawn from a data processing department is presented to demonstrate the role of QI stories in an organization's improvement efforts.* The QI story is presented in Figure 4.23. This QI story goes through two iterations of the PDSA cycle; however, a never-ending set of PDSA iterations will follow as the data processing department pursues continuous improvement in its daily work. The first iteration of the PDSA cycle focuses attention on all data entry operators in the data processing department. In this iteration of the PDSA cycle, selecting a theme is presented in QI story board 1. This includes showing the background of theme selection and the reason for selecting the theme in relation to the organization's and department's objectives. An understanding of the present situation is presented in QI story board 2. An analysis of the present situation, which is shown in QI story board 3, is performed to determine an appropriate plan that pursues the theme and the organization and department objectives. Setting the plan into motion on a trial basis is presented in QI story board 4. The effectiveness of the plan is measured on the theme, and the organization and department objectives. This is shown in QI story board 5. Standard operating procedure is set, which formalizes the countermeasures and prevents backsliding. This is shown in QI story board 6. A plan for future actions is in QI story board 7.

The second iteration of the PDSA cycle focuses attention on an individual data entry operator. In this iteration, selecting a theme is accomplished when the data processing manager realizes that future process improvements require her to identify and improve operators whose error rate is out of control on the high side (see QI story board 8). An understanding of the present situation determined that data entry operators 004 and 009 were out of control on the high side. This is presented in QI story board 9. An analysis of the present situation, which is shown in QI story board 10, determined the actions necessary to improve operator 004. The manager put the plan into action. This is shown in QI story board 11. The positive effect of the plan on operator 004 and the organization and department objective was confirmed. This is shown in QI story board 12. Standard operating procedure was set and formalized the actions of all operators to prevent backsliding. This is presented in

---

* See Gitlow, Oppenheim, and Oppenheim, *Quality Management*, pp. 388–403.

QI story board 13. Finally, a plan for future actions is specified in QI story board 14.

Empowerment

Steps 18 and 19 of the Detailed Fork Model in Figure 1.2 include empowering employees through daily management.* "Empowerment" is a term commonly used by managers in today's organizational environment. However, empowerment has not been operationally defined, and its definition varies from application to application. Currently, the popular definition of empowerment relies loosely on the notion of dropping decision-making

QI STORY BOARD 1

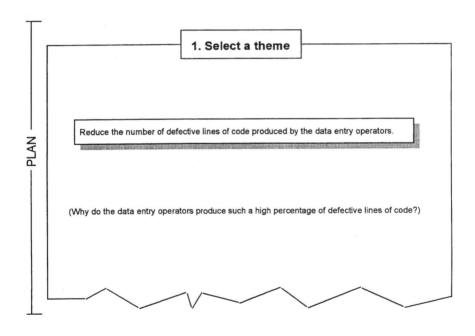

Figure 4.23 Quality Improvement (QI) Story Boards 1 through 14

---

* Modified from Pietenpol, D. and Gitlow, H., "Empowerment and the System of Profound Knowledge," *International Journal of Quality Science*, vol. 1, no. 3, 1996, pp. 50–57.

QI STORY BOARD 1 (con't)

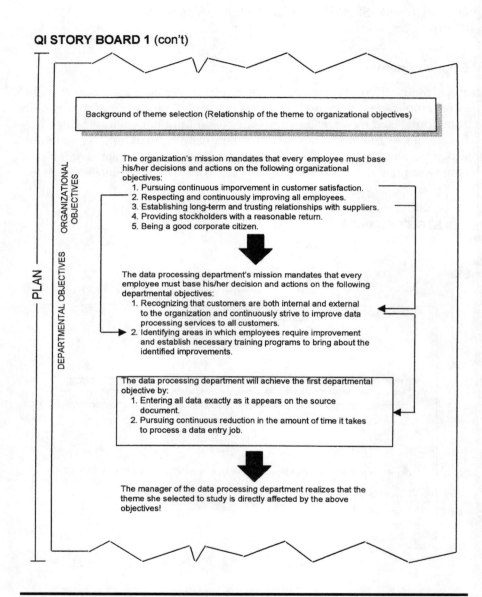

Background of theme selection (Relationship of the theme to organizational objectives)

The organization's mission mandates that every employee must base his/her decisions and actions on the following organizational objectives:
1. Pursuing continuous imporvement in customer satisfaction.
2. Respecting and continuously improving all employees.
3. Establishing long-term and trusting relationships with suppliers.
4. Providing stockholders with a reasonable return.
5. Being a good corporate citizen.

The data processing department's mission mandates that every employee must base his/her decision and actions on the following departmental objectives:
1. Recognizing that customers are both internal and external to the organization and continuously strive to improve data processing services to all customers.
2. Identifying areas in which employees require improvement and establish necessary training programs to bring about the identified improvements.

The data processing department will achieve the first departmental objective by:
1. Entering all data exactly as it appears on the source document.
2. Pursuing continuous reduction in the amount of time it takes to process a data entry job.

The manager of the data processing department realizes that the theme she selected to study is directly affected by the above objectives!

(left margin labels, top to bottom) ORGANIZATIONAL OBJECTIVES / DEPARTMENTAL OBJECTIVES / PLAN

Figure 4.23　Quality Improvement (QI) Story Boards 1 through 14

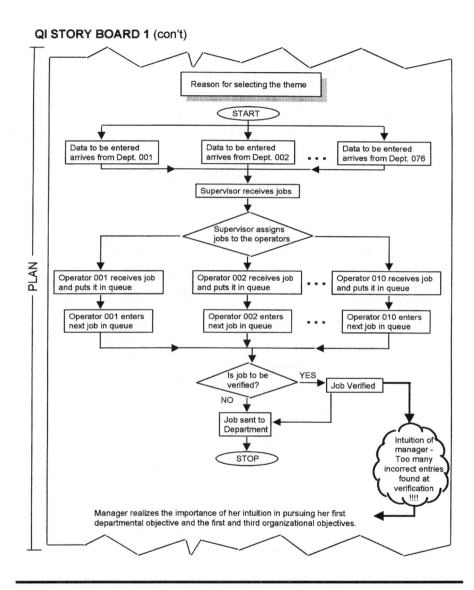

Figure 4.23 Quality Improvement (QI) Story Boards 1 through 14

QI STORY BOARD 2

2. Grasp of the present situation

Manager's intuition leads her to conduct a survey to determine customer (other departments) satisfaction with her department's performance.

Manager constructs a list of her department's customers.

Administration
Production
Marketing
.
.
.

Manager constructs a questionnaire to determine customer satisfaction.

PLAN

Department: _____
Supervisor: _____

(1) Do you feel that the error rate that data entry provides your department is:
unsatisfactory [ ]   satisfactory [ ]   excellent [ ]

(2) Approximately what percent of the data entry errors your department receives from our department contain errors attributable to our department? _____%

Figure 4.23 Quality Improvement (QI) Story Boards 1 through 14

QI STORY BOARD 2 (con't)

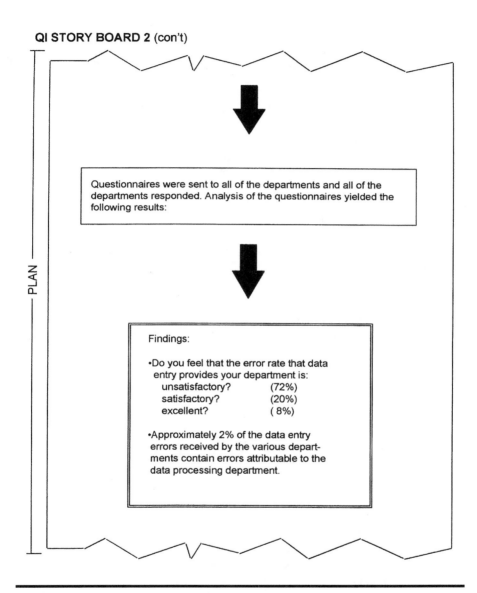

Questionnaires were sent to all of the departments and all of the departments responded. Analysis of the questionnaires yielded the following results:

Findings:

•Do you feel that the error rate that data entry provides your department is:

unsatisfactory?	(72%)
satisfactory?	(20%)
excellent?	(8%)

•Approximately 2% of the data entry errors received by the various depart- ments contain errors attributable to the data processing department.

Figure 4.23 Quality Improvement (QI) Story Boards 1 through 14

QI STORY BOARD 2 (con't)

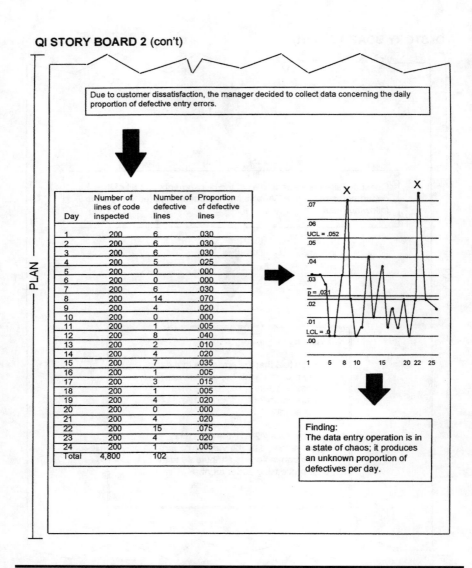

Due to customer dissatisfaction, the manager decided to collect data concerning the daily proportion of defective entry errors.

PLAN

Day	Number of lines of code inspected	Number of defective lines	Proportion of defective lines
1	200	6	.030
2	200	6	.030
3	200	6	.030
4	200	5	.025
5	200	0	.000
6	200	0	.000
7	200	6	.030
8	200	14	.070
9	200	4	.020
10	200	0	.000
11	200	1	.005
12	200	8	.040
13	200	2	.010
14	200	4	.020
15	200	7	.035
16	200	1	.005
17	200	3	.015
18	200	1	.005
19	200	4	.020
20	200	0	.000
21	200	4	.020
22	200	15	.075
23	200	4	.020
24	200	1	.005
Total	4,800	102	

Finding:
The data entry operation is in a state of chaos; it produces an unknown proportion of defectives per day.

Figure 4.23 Quality Improvement (QI) Story Boards 1 through 14

QI STORY BOARD 3

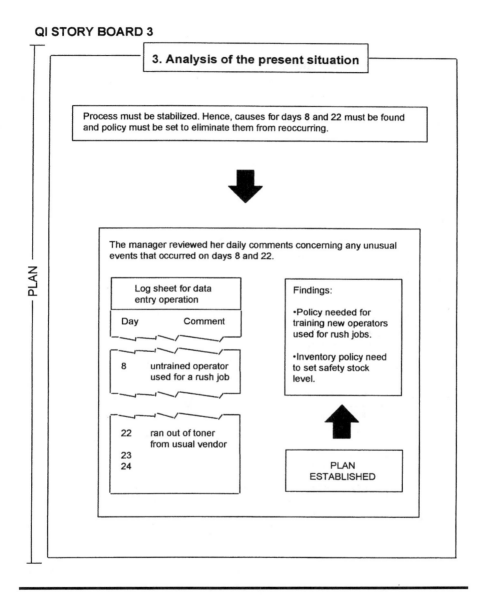

Figure 4.23 Quality Improvement (QI) Story Boards 1 through 14

QI STORY BOARD 4

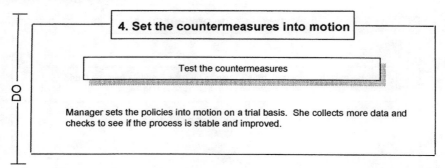

QI STORY BOARD 5

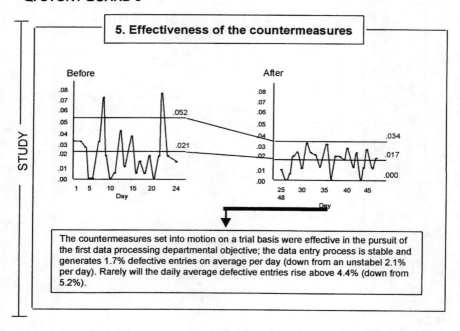

Figure 4.23 Quality Improvement (QI) Story Boards 1 through 14

QI STORY BOARD 6

6. Standard operating procedure

The manager establishes formal operating procedures, including appropriate training, for inventory policy and new operator skills development.

The manager decides that a random sample of 200 lines per month will be drawn from every operator's output. These samples will be analyzed so that appropriate actions can be taken to prevent any backsliding in areas that have been improved.

ACT

QI STORY BOARD 7

7. Plan for future actions

		When will future plans be carried out										Who will carry out plan
		12-87	1-88	2-88	3-88	4-88	5-88	6-88	7-88	8-88	9-88	
Phase 1	Work with operator 004	◄──────►										Manager and 004
Phase 2	Work with operator 009						◄──►					Manager and 009
Phase 3	Check progress of entire department								◄──►			Manager and 001-010
Phase 4	Survey customers to determine satisfaction with k.p										──►	Manager

ACT

Figure 4.23 Quality Improvement (QI) Story Boards 1 through 14

QI STORY BOARD 8

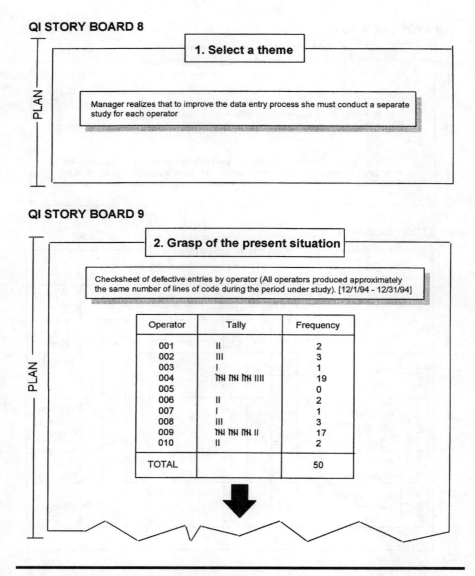

QI STORY BOARD 9

Figure 4.23 Quality Improvement (QI) Story Boards 1 through 14

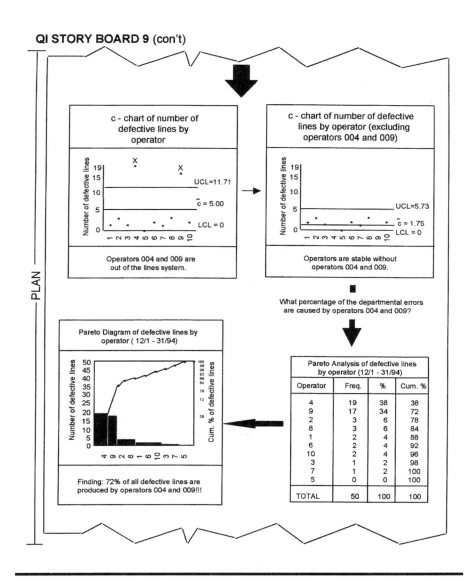

Figure 4.23 Quality Improvement (QI) Story Boards 1 through 14

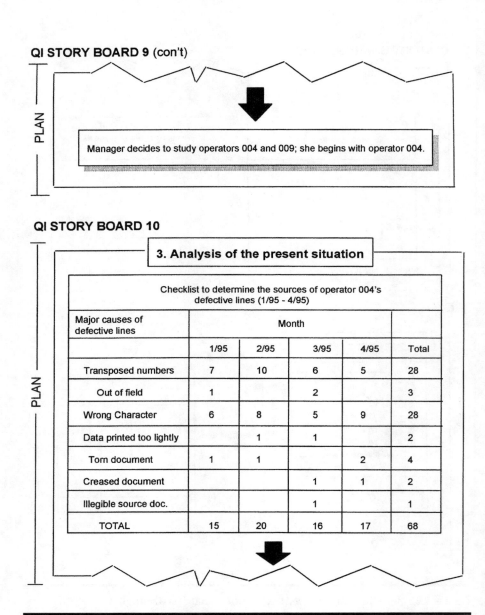

QI STORY BOARD 9 (con't)

Manager decides to study operators 004 and 009; she begins with operator 004.

QI STORY BOARD 10

3. Analysis of the present situation

Checklist to determine the sources of operator 004's defective lines (1/95 - 4/95)

Major causes of defective lines	Month				
	1/95	2/95	3/95	4/95	Total
Transposed numbers	7	10	6	5	28
Out of field	1		2		3
Wrong Character	6	8	5	9	28
Data printed too lightly		1	1		2
Torn document	1	1		2	4
Creased document			1	1	2
Illegible source doc.			1		1
TOTAL	15	20	16	17	68

Figure 4.23 Quality Improvement (QI) Story Boards 1 through 14

QI STORY BOARD 10 (con't)

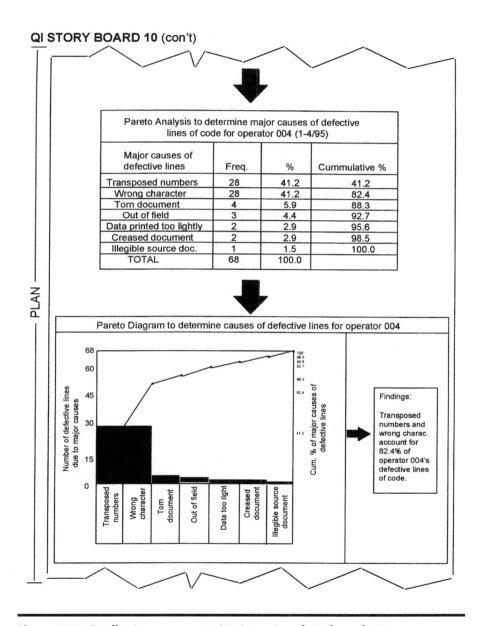

Pareto Analysis to determine major causes of defective lines of code for operator 004 (1-4/95)

Major causes of defective lines	Freq.	%	Cummulative %
Transposed numbers	28	41.2	41.2
Wrong character	28	41.2	82.4
Torn document	4	5.9	88.3
Out of field	3	4.4	92.7
Data printed too lightly	2	2.9	95.6
Creased document	2	2.9	98.5
Illegible source doc.	1	1.5	100.0
TOTAL	68	100.0	

Pareto Diagram to determine causes of defective lines for operator 004

Findings:

Transposed numbers and wrong charac. account for 82.4% of operator 004's defective lines of code.

PLAN

Figure 4.23 Quality Improvement (QI) Story Boards 1 through 14

QI STORY BOARD 10 (con't)

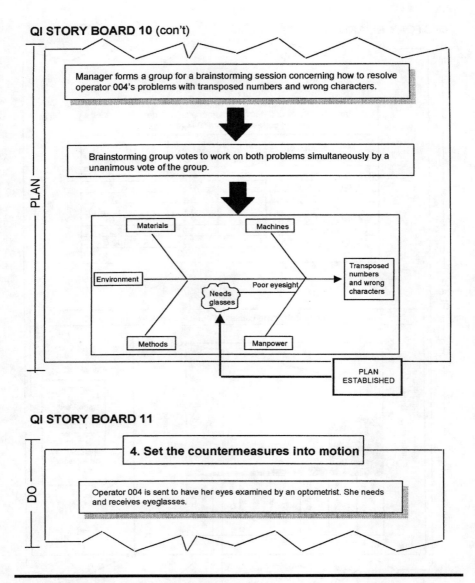

QI STORY BOARD 11

Figure 4.23 Quality Improvement (QI) Story Boards 1 through 14

QI STORY BOARD 12

5. Effectiveness of the countermeasures

Manager collects 25 additional daily samples of 200 lines of code each to determine the effect of operator 004's eyeglasses on her defective rate.

Day	Number Defective	Total Lines of Code	Proportion Defective
1	2	200	0.010
2	3	200	0.015
3	2	200	0.010
.	.	.	.
.	.	.	.
.	.	.	.
25	2	200	0.010
	40	5000	0.008

p charts comparing the proportion of defective lines produced by the "average" operator before improvement efforts with the proportion of defective lines produced by operator 004 after improvement efforts.

Finding: Operator 004 is stable and producing 8 defective lines per 1,000. Rarely will her defect rate go above 2.6 per 1,000. The countermeasure taken with operator 004 is effective in the pursuit of the first data processing department objective.

STUDY

Figure 4.23 Quality Improvement (QI) Story Boards 1 through 14

QI STORY BOARD 13

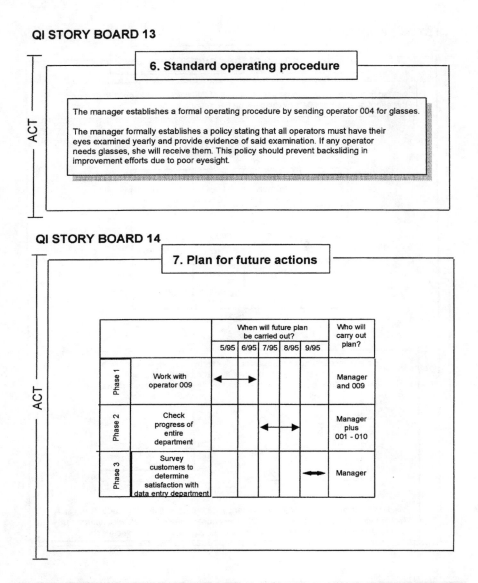

6. Standard operating procedure

The manager establishes a formal operating procedure by sending operator 004 for glasses.

The manager formally establishes a policy stating that all operators must have their eyes examined yearly and provide evidence of said examination. If any operator needs glasses, she will receive them. This policy should prevent backsliding in improvement efforts due to poor eyesight.

QI STORY BOARD 14

7. Plan for future actions

		When will future plan be carried out?					Who will carry out plan?
		5/95	6/95	7/95	8/95	9/95	
Phase 1	Work with operator 009	◄──►					Manager and 009
Phase 2	Check progress of entire department			◄──►			Manager plus 001 - 010
Phase 3	Survey customers to determine satisfaction with data entry department					◄─►	Manager

Figure 4.23 Quality Improvement (QI) Story Boards 1 through 14

down to the lowest appropriate level in an organization. The basic premise of empowerment is that if people are given the authority to make decisions, they will take pride in their work, be willing to take risks, and work harder to make things happen. Frequently, the reality of empowerment is that employees are empowered until they make a mistake; then the hatchet falls. Most employees know this and treat the popular concept of empowerment with the lack of respect it merits. Empowerment in its current form is destructive to quality management.

Empowerment in a Quality Management sense has a dramatically different aim and definition. The aim of empowerment in Quality Management is to increase pride in work and joy in the outcome for all employees.

The definition of empowerment that translates the above aim into a realistic objective follows. *Empowerment* is a process that provides an individual or group of employees the opportunity to:

1. Define and document methods.
2. Learn about methods through training and development.
3. Improve and innovate best practice methods that make up systems.
4. Utilize latitude in their own judgment to make decisions within the context of best practice methods.
5. Trust superiors to react positively to the latitude taken by employees making decisions within the context of best practice methods.

Empowerment starts with leadership, but requires the commitment of all employees. Leaders provide employees with all five items stated above. Employees accept responsibility for:

1. Increasing their training and knowledge of methods and the systems of which they are a part.
2. Participating in the development, standardization, improvement, and innovation of best practice methods.
3. Increasing their latitude in decision-making within the context of best practice methods.

Latitude to make decisions within the context of a best practice method refers to the options an employee has in resolving a problem within the confines of a best practice method, not the modification of the best practice method. Differentiating between the need to change the best practice method and latitude within the context of the best practice method takes place at the operational level.

Empowerment can exist only in an environment of trust that supports planned experimentation concerning ideas to improve and innovate best practice methods. Ideas for improvement and innovation can come from individuals or from the team, but tests of the worthiness of those ideas are conducted through planned experiments under the auspices of the team (the "Do" stage of the PDSA cycle). Anything else will result in chaos, because everybody will do his or her own thing.

Individual employees are taught to understand that increased variability in output will result if each employee follows his or her own method. This increased variability will create additional cost and unpredictable customer service. Employees are educated about the need to reach consensus on one "best practice" method.

The "best practice" method will consist of generalized procedures and individualized procedures. Generalized procedures are standardized procedures that all employees follow. Individualized procedures are procedures that afford each worker the opportunity to utilize their individual differences by creating their own standardized procedure. However, the outputs of individualized procedures are standardized across employees. Individualized procedures can be improved through personal efforts. In the beginning of a quality improvement effort, employees and management may not have the knowledge to allow for individualized procedures.

A professor following an approved departmental syllabus for a certain course is an example of an employee using a generalized procedure. When that professor injects his or her own stories, examples, and jokes, he or she is using individual procedures.

Empowerment is operationalized at two levels. First, employees are empowered to develop and document best practice methods using the SDSA cycle. Second, employees are empowered to improve or innovate best practice methods through application of the PDSA cycle.

Applying PDSA to Daily Management

As managers see the results from improved processes, they will want to expand the number of daily management project teams. This should be discouraged in the beginning of a Quality Management effort.

Instead, managers should be asked to direct their existing project teams to continually apply PDSA to the processes already under study. The benefit of this action is ensuring that managers learn how to continuously improve and innovate processes, not how to make one improvement in a process and jump to another process. Management reviews are an excellent vehicle to promote this type of training experience. Reviewers can ask the following question: "Can I see your improvement action memoranda for this process?" The reviewee should be able to show multiple improvement action memoranda, including changes to training programs, for the process they are studying with their project team.

Coordinating Project Teams

As the initial process improvement teams begin to show positive results, other process improvement teams will be formed by area or department managers in response to localized issues (see step 19 of the Detailed Fork Model in Figure 1.2).

The initial and other process improvement teams require resources, such as PILs, members to work on projects, training, financial resources, physical space in which to meet, and the direction and guidance of a higher level of management.

As the number of teams increases, a structure to coordinate and manage the teams at the department level is necessary. The structure is called a Local Steering Team (LST) (see step 20 of the Detailed Fork Model in Figure 1.2). Each department's LST has the responsibility to coordinate daily management projects (see steps 18 and 19 of the Detailed Fork Model in Figure 1.2).

SUMMARY

Chapter 4 presents a discussion of Prong One of the fork model, daily management. Daily management is used to develop, maintain, improve, and innovate the methods employed in daily work.

The first phase of implementing daily management involves selecting initial project teams. Process Improvement Leaders are chosen by the EC and trained in "Basic Quality Control Tools" and "Psychology of the Individual and Team." Then the initial projects are determined, and project teams are formed.

After training, each initial project team works on one or more methods using daily management. Daily management includes housekeeping, which is the development, standardization, and deployment of methods required for daily work, and daily management, which is the maintenance, improvement, and innovation of methods for daily work.

Housekeeping functions are developed through function deployment. This is the way employees determine what functions are required to perform each method they use in their daily work. Housekeeping is accomplished by employing the SDSA cycle. The objective is to determine the "best practice method" for each function. Best practice methods are monitored through measurements that are operationally defined and measure results and the processes that generate results.

After a best practice method has been developed and deployed by a project team, daily management activities begin. Daily management is used to reduce process variation and to center the process on the customer's requirements. The PDSA cycle is utilized in daily management in a continuous progression of never-ending improvement. A personal example of using daily management to improve a "best practice method" for exercise habits is presented in this chapter. Also, a business example of using daily management to develop and improve a "best practice method" for the wettability of the surface of cuvettes, using cold gas plasma treatment, is presented in this chapter.

Project teams present their housekeeping and daily management projects to managers in management reviews. This is a process that involves comparing the actual results generated with the targets established. Suggested questions that can form the basis of a management review are listed in this chapter. It is critical that management reviews take into account common and special causes of variation. If a management review is done properly, there is no place for tampering with the process or blaming employees for problems out of their control.

A special procedure called the Quality Improvement (QI) story is used to present and deal with pressing daily problems that cannot be adequately handled through regular reviews. QI stories standardize the reporting of process improvement efforts, help avoid logical errors in analysis, and make process improvement efforts easier to deploy company-wide. The seven-step procedure for constructing a QI story is described in this chapter. A QI story drawn from a data processing department is presented to show the role of QI stories in an organization's improvement efforts.

Empowering employees through daily management is discussed in this chapter. The aim of empowerment in a quality management sense is to increase pride in work and joy in the outcome for all employees. It is operationalized at two levels. First, employees are empowered to develop and document best practice methods using the SDSA cycle. Then they are empowered to improve or innovate best practice methods through the application of the PDSA cycle.

As the initial project teams begin to show positive results, and more teams are formed by area or department managers, a structure is needed to coordinate the teams at the departmental level. This structure is called the Local Steering Team.

5

PRONG TWO: CROSS-FUNCTIONAL MANAGEMENT

PURPOSE OF THIS CHAPTER

The purpose of this chapter is to explain what is required to develop, standardize, deploy, maintain, improve, and innovate methods that cross areas in an organization. Cross-functional management is Prong Two of the Quality Management model presented in this book.

BACKGROUND

Cross-functional management* is critical to the Quality Management model because it weaves together the vertical (line) functions of management with the horizontal (interdepartmental) functions of management. Professor Kaoru Ishikawa states that "in order to be called a fabric, both horizontal and vertical threads need to be woven together, and only when horizontal or Cross-functional management threads are woven together with vertical threads can a company be considered similarly cohesive."** Cross-functional management is important because it promotes the reorganization of corporate management systems to improve interdepartmental communication and cooperation, and provides clear lines of responsibility for that reorganization.

---

* This chapter draws heavily from Kurogane, K., *Cross-Functional Management: Principles and Practical Applications*, Asian Productivity Organization (Tokyo), 1993, pp. 33–36.
** Ishikawa, K., "Management in Vertical-Threaded Society," *Quality Control*, vol. 32, no. 1, 1981, pp. 4–5.

The primary areas for the application of cross-functional management include quality management (quality control and quality improvement), cost management (profit management and cost reduction), delivery management (production quantity management, delivery date management, and production system management), and personnel management (human development, education, and work morale enhancement). Quality and cost are usually the first areas to receive attention in cross-functional management.

The auxiliary areas for the application of cross-functional management include new product development (research and development, technology development, and production technology), sales management (marketing, sales activity management, and sales expansion), safety management (safety/hygiene control, labor safety control, and environmental control), and QC promotional support (QC circle standardization). Primary cross-functional areas are permanent. Auxiliary cross-functional areas change according to current and expected conditions.

SELECTING INITIAL CROSS-FUNCTIONAL TEAMS

The members of the EC form initial cross-functional teams (see step 21 of the Detailed Fork Model in Figure 1.2), usually in the areas of quality or cost management. The EC selects a leader for each team (see step 22 of the Detailed Fork Model in Figure 1.2) and allocates appropriate resources for the education and training of the leader (see step 23 of the Detailed Fork Model in Figure 1.2). Each cross-functional leader is an executive with the title of Senior Vice President or Vice President in charge of a function. The EC uses the recommendations of the team leader to select members for the initial cross-functional teams (see step 24 of the Detailed Fork Model in Figure 1.2). Team size is kept to a minimum, about five people. All team members are trained in appropriate theory and practice (see step 24 of the Detailed Fork Model in Figure 1.2). Team members should be executives with the rank of director or above. All team members do not have to come from affected areas. A diversity of opinion and knowledge is helpful, but it is not necessary to have all affected areas represented on a cross-functional team. The team facilitator is an executive in charge of a function, such as personnel. The support staff for a cross-functional team is from the team leader's home department because the team leader needs to have the authority to make things happen for his or her cross-functional team.*

---

* Kurogane, *Cross-Functional Management,* p. 45.

DOING CROSS-FUNCTIONAL MANAGEMENT

Cross-functional management includes the following activities:

1. Studying and applying Dr. Deming's theory of management to company-wide systems.
2. Developing measurements for company-wide systems.
3. Coordinating and optimizing company-wide systems within departmental methods.
4. Allocating resources for cross-functional and departmental methods by establishing targets.
5. Ensuring that each department performs its deployed methods in daily management.
6. Monitoring company-wide systems in respect to targets from a corporate level (management review).
7. If necessary, taking action utilizing the PDSA cycle to decrease the difference between actual results and targets (variance analysis).

The following steps are used to do cross-functional management:*

1. Clarify the purpose (aim) of the cross-functional management effort. Is the purpose of cross-functional management to deploy a cross-functional process into a division or department to achieve a rational divisional or departmental target? Quality improvement and cost reduction are examples of cross-functional processes deployed into divisional or departmental processes to attain a rational divisional or departmental target. Or, is the purpose of cross-functional management to improve a divisional or departmental process that effects areas across the organization to achieve a rational divisional or departmental target? R&D and human resources are examples of divisional or departmental processes that affect areas across an organization and must be improved to attain a rational target.
2. Prepare a list of the divisions or departments that will participate in the proposed cross-functional team.
3. Construct an integrated flowchart of the proposed process. For quality and cost, the integrated flowchart is a flowchart arranged in a matrix format with stakeholders, tools, and documents of the process as the columns and steps of the proposed process under study in the rows.

---

* Modified from Mizuno, S., *Company-Wide Total Quality Control*, Asian Productivity Organization (Tokyo), 1988, p. 108.

The listing of each step in the proposed process should include the required activities and needed items, the individual or group responsible to perform the activity, and the persons responsible for the results of the activity. The cells of the matrix indicate the relationships between stakeholders, tools and documents, and the steps of the proposed process. An example of an integrated flowchart for a personnel management process appears in Figure 5.1.

4. Create measures to monitor the existing and proposed processes.
5. Identify and select priority projects using the proposed process and measures developed by team members. Application of the System of Profound Knowledge is the key to effective cross-functional management. In quality, such a project might reduce the number of customer complaints. In cost, such a project might decrease the cost to manufacture a product.

As expertise is developed with a company-wide system, it is deployed into daily management methods where appropriate (see step 25 of the Detailed Fork Model in Figure 1.2).

STRUCTURES FOR CROSS-FUNCTIONAL MANAGEMENT

Several administrative structures can be used to promote cross-functional management. In small organizations, one cross-functional team comprising all relevant executives can be established to coordinate and optimize company-wide systems. In large organizations, one cross-functional team comprising appropriate executives can be established to coordinate and optimize each company-wide system. For example, there could be one team for quality management, one team for safety/hygiene management, and so on. Another alternative for large organizations is to allow a functional department to coordinate and optimize one company-wide system. For example, the Personnel Department could coordinate and optimize the company-wide systems dealing with the enhancement of employee morale.

Frequently, executives claim that they do not have time for cross-functional management due to the demands of their daily routine. It is necessary for these executives to do daily management to remove non-value-added daily routine from their schedules to free up time for cross-functional management.

Cross-functional teams report directly to the members of the EC and have the highest level of decision-making authority. They perform the Plan and Study phases of the PDSA cycle for company-wide systems. Implementation of company-wide systems, the Do and Act phases of the PDSA cycle, is carried out by line departments in daily management.

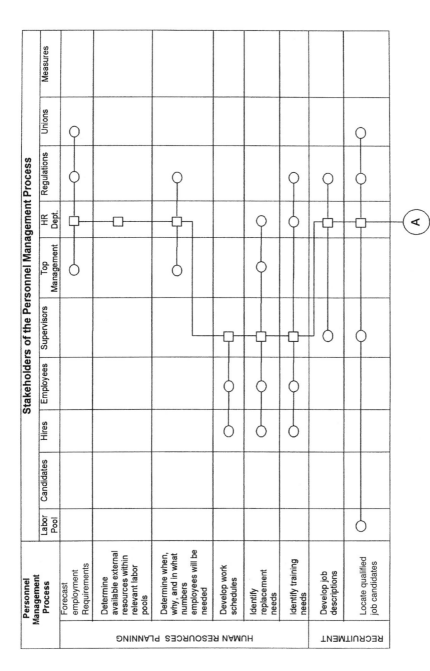

Figure 5.1 Integrated Flowchart of the Personnel Management Process

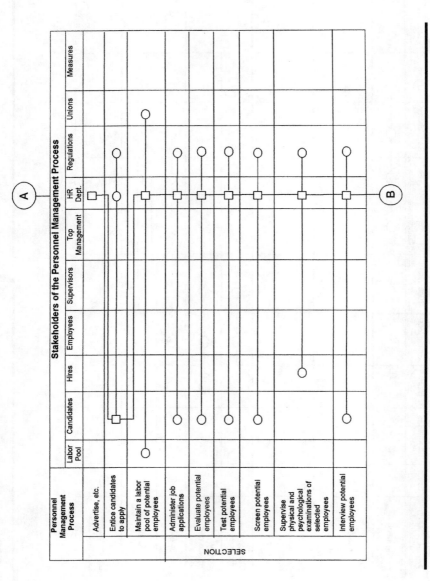

Figure 5.1 (continued) Integrated Flowchart of the Personnel Management Process

Stakeholders of the Personnel Management Process

Personnel Management Process	Labor Pool	Candidates	Hires	Employees	Supervisors	Top Management	HR Dept.	Regulations	Unions	Measures
Select finalists		O					□			
Send finalists to hiring department		O					□			
Interview potential finalists		O			O					
Select hiree		O			□					
Hire employee			O		□	O	□	O	O	
File paperwork						O	□	O		
Familiarize employees with company policy			O				□	O		
Familiarize employees with safety codes			O				□	O		
Familiarize employees with safety codes			O		O		□	O		
Familiarize employees with work expectations			O		O		□	O		
If appropriate, provide technical training in specific work conditions			O		O		□	O		

ORIENTATION

Figure 5.1 (continued) Integrated Flowchart of the Personnel Management Process

Stakeholders of the Personnel Management Process

Personnel Management Process		Labor Pool	Candidates	Hires	Employees	Supervisors	Top Management	HR Dept.	Regulations	Unions	Measures
ORIENTATION	If appropriate, provide technical training in equipment			O		O		□	O		
	If appropriate, provide technical training in processes			O		O		□	O		
TRAINING (VOCATIONAL SKILLS)	On-job training			O		O		□	O	O	
	Off-job training (e.g. public seminars)			O		O		□	O	O	
	Vestibule training (e.g. practice at work site)			O		O		□	O	O	
	Institutional training (e.g. corporate university)			O		O		□	O	O	
DEV'T.	Job enhancement				O	□		O	O	O	
	Job advancement				O	□		O	O		
COMPENSATION MANAGEMENT	Wage and salary determination					□	O	□	O	O	
	Raises				O	□	□	O	O	O	
	Bonuses				O		□	O	O	O	
	Other monetary issues						O	□	O	O	

C — D

Figure 5.1 (continued) Integrated Flowchart of the Personnel Management Process

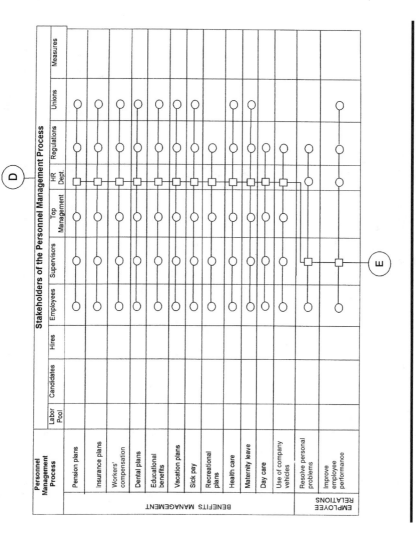

Figure 5.1 (continued) Integrated Flowchart of the Personnel Management Process

Stakeholders of the Personnel Management Process

Personnel Management Process		Labor Pool	Candidates	Hires	Employees	Supervisors	Top Management	HR Dept.	Regulations	Unions	Measures
PERFORMANCE	Safety				O	□		O	O	O	
	Health				O	□		O	O	O	
	Collective bargaining					O	O	□	O	O	
	Relationships between management and legally constituted employee unions and associations						O	□		O	
PERFORMANCE	Appraise subordinate's behavior				O	□		O		O	
	Provide feedback for improvement				O	□		O		O	
TRANSFERS	Change in employee's job				O	O	O	□	O	O	
	Change in employee's position (promotion or demotion)				O	O	O	□	O	O	
TERMINATIONS	Quit				□	O	O	O	O	O	
	Fired				O	□	O	O	O	O	
	Retired				□	O	O	O	O	O	
	Death				□	O	O	O	O	O	
	Layoff				O	O	□	O	O	O	

Figure 5.1 (continued) Integrated Flowchart of the Personnel Management Process

COORDINATING CROSS-FUNCTIONAL TEAMS

As the initial cross-functional teams successfully improve company-wide systems, the EC will form new cross-functional teams (see step 26 of the Detailed Fork Model in Figure 1.2). The EC reviews, manages, and coordinates all cross-functional teams (see step 27 of the Detailed Fork Model in Figure 1.2).

A cross-functional management review of the line departments affected by cross-functional policy is conducted by a cross-functional team leader one or more times per year. These reviews study departmental management from a company-wide perspective, and provide feedback to line departments and the cross-functional team for the next year. Line departments report their progress with implementing cross-functional policy by filing a cross-functional management report. The cross-functional team collects all departmental cross-functional management reports and uses them as a basis for conducting reviews and taking action. Furthermore, the cross-functional management team reports its findings to the EC.

Cross-functional teams generate projects that may be sent to the Policy Deployment Committee (see step 29 of the Detailed Fork Model in Figure 1.2, to be discussed in the next chapter) or a Local Steering Team for action (see step 20 of the Detailed Fork Model in Figure 1.2).

SOME COMMON PROBLEMS IN IMPLEMENTING CROSS-FUNCTIONAL MANAGEMENT

Cross-functional activities, due to their interdisciplinary structure, are activities that are ripe for misunderstandings between team members, and between team members and the rest of the organization. For example, a cross-functional team working on the budgeting and planning process can easily create confusion, resentment, and fear among the members of an organization. This happens when the cross-functional team changes the methods for allocating resources to departments and thereby reduces a department's ability in the short term to predict its budget line.

Some common mistakes made when cross-functional teams are established are discussed below. The mistakes cover the longevity, membership, focus, resources, and communications of cross-functional teams.

Cross-functional teams are permanent committees that deal with the continuous improvement of important company-wide systems over the long term. Dissolving a cross-functional team after its members have solved some problem in a company-wide system is not advisable. For example, a cross-functional team would be established to improve the company-wide safety/hygiene system over the long term, rather than created to deal with a rash of industrial accidents in the short term.

Cross-functional teams don't have to include representatives from all areas affected by their policy. Including members from all areas on a cross-functional team may make the team too big to manage well. For example, a cross-functional team that addresses cost management does not have to include representatives from all areas in an organization.

Cross-functional team members transcend the boundaries of their own areas. A person from the production area learns to think in terms of the entire system of interdependent stakeholders when addressing company-wide systems, not from the perspective of the production area. People on cross-functional teams who represent their own special interest groups, not the welfare of the entire organization, are not ready to participate on a cross-functional team. They need further training.

It is extremely important that cross-functional policy be communicated to all relevant members of an organization's interdependent system of stakeholders. Only through communication can people understand and buy into the company-wide changes that can emanate from a cross-functional team. Recall from step 10 of the Detailed Fork Model (in Chapter 2) that diffusion of a new idea, in this case a cross-functional policy, requires a specific plan of action that is based on the appropriate theories of communication.

A GENERIC EXAMPLE OF CROSS-FUNCTIONAL MANAGEMENT: STANDARDIZATION OF A CORPORATE-WIDE METHOD FOR CUTTING COST

Purpose

This section demonstrates how cross-functional management utilizes the System of Profound Knowledge to create new processes to resolve existing problems in corporate-wide processes. The corporate-wide process examined in this section is the cost-cutting process.

Traditional Cost-Cutting Process

A common process for cutting costs x% or $$y$ consists of three steps. First, managers build up a layer of inefficiency over time. For example, if a job becomes obsolete due to computerization, the person holding the job is transferred to other work or let go, but the job is not deleted from the payroll. If the manager is asked to cut cost by x% or $$y$, he or she puts forth the hidden resources from the obsolete job. Second, managers identify nonessential expense items. For example, in some departments training dollars or gifts/awards dollars or travel dollars are nonessential to the functioning of the department. If the manager cannot cut his or her costs x% or y using the method in the first step, he or she will cut nonessential expense items in the budget, in part or in whole. Third, managers prioritize essential expense items for budget cuts. For example, managers may rank their personnel from most meritorious to the least meritorious, or most senior to most junior. If the manager cannot cut his or her costs x% of y using the methods in the first and second step, he or she will begin to cut essential expense items until the required x% or y cost cut is achieved in his or her area.

The above process for cutting costs causes managers to hide resources from top management. Frequently, the hidden resources are desperately needed elsewhere in the organization. This results in suboptimization of the organization as a whole.

Cross-Functional Management Cost-Cutting Process

A cross-functional management cost-cutting process is used by a manager to attain a rational x% or y decrease in costs using process improvement in his or her area. The following steps explain how a manager applies a cross-functional management process in his or her daily management.

1. Clarify the purpose (aim) of the proposed daily management process in terms of a x% or y cut in costs. The x% or y cut in costs should not be an arbitrary numerical goal; rather, it should be based on data (e.g., breakeven analysis).
2. Prepare a list of the divisions or departments that are affected by the proposed daily management process.
3. Construct an integrated flowchart of the proposed daily management process that reflects the x% or y cut in costs.
4. Identify measures to monitor the existing and proposed daily management process. Show actual documentation on costs, revenues, etc. for the existing and proposed daily management process.
5. Implement the proposed process and measures. Show documentation on costs, revenues, etc. for the proposed process.

If the proposed daily management process achieves the x% or y cost cut, team members continue to improve the process. If the proposed daily management process does not achieve the cost-cutting goal, team members continue to try with the help of higher management. If this works, team members continue to improve the process.

If the proposed daily management process does not achieve the *x*% or $y cost cut with the help of higher management, then top management makes decisions about the reallocation of organizational resources. If this works, it is a temporary measure and team members continue to try to cut costs, with the assistance of top management or outside expertise, if needed.

APPLICATION OF THE COST-CUTTING PROCESS IN HUMAN RESOURCES

This section shows an application of the above cross-functional management cost-cutting process to the Selection subsystem of the Personnel Management process in Figure 5.1. The aim of the application is to cut costs in the Human Resources department by $100,000. The $100,000 figure is based on analyses of industry practices and the cost structure of all departments in the organization. Top management has determined that the Human Resource department needs to cut costs by $100,000 to optimize the entire organization. Other departments' cost cuts range between $0 and $1,500,000. The $100,000 cost cut is not an arbitrary numerical goal.

A list of the stakeholders of the Selection subprocess of the Personnel Management process includes new hires, employees, supervisors, top managers, and the employees of the Human Resources department. The director of the Human Resources department will chair this team.

An integrated flowchart of the existing Selection subprocess can be seen in the *"Selection"* section of Figure 5.1. The key measure used to monitor the effect of change on the Selection sub-process is the number of screening examinations (scholastic verification, crime check, and drug test) per month [EX]. Baseline data will be collected for the measure before any process change is put into practice. Also, data will be collected for the measure after any process change is put into practice. The relevant costs for the existing cost-cutting process include scholastic verification costs ($15.50 per candidate), crime check costs ($24.50 per candidate), and drug test costs ($84.00 per candidate). If there are *n* candidates who are screened in a month, then the monthly screening cost is $124*n*.

An improvement to the existing Selection subprocess shown in Figure 5.1 follows. It is to move the *"Supervise physical and psychological examinations of selected employees"* step from its current position in the selection process after *"Screen potential employees"* and before *"Interview potential employees"* to a position after the *"Select hire"* step and before the *"Hire employee"* step. The logic of this change is that a very low percentage of candidates fail the screening examinations, hence, screening examinations do not effectively reduce the candidate pool. Screening examinations should be done only on the finalist candidate for a given job, to minimize costs.

The relevant costs for the improved Selection subprocess are one set of screening examinations for each finalist, for each job. If a finalist passes the screening examinations, then the total screening cost for any given job is $124.

In conclusion, the Human Resources department used to conduct about 84 screening examinations per month (approximately 1,000 per year), at a yearly cost of $124,000. Now, the Human Resources department conducts screening examinations only on finalists. For the same period of time, there were 100 finalists. The screening costs in the new process are $12,400, assuming all finalists pass their screening examinations. This represents a savings of $111,600 ($124,000 − $12,400), which surpasses the needed cost cut of $100,000.

A MANUFACTURING APPLICATION: TOYOTA FORKLIFT*

Background

This section presents the cross-functional management quality assurance process for new product development used in all six divisions of the Toyota Motor Corp. The process was developed using the five step cross-functional management model presented in the beginning of this chapter. Additionally, this section includes an application of the model in the development of the Toyota X300 forklift.

Each division of Toyota is responsible for the development of new products. The process begins with input from the long-range business plan and annual policy statement of a division. The departments in each division responsible for new product development assume development tasks based on Toyota's cross-functional quality assurance process (Figure 5.2). These responsibilities span product planning through production preparation.

Management reviews are conducted by managers to check and follow-up new product development at predetermined intervals. Design reviews are conducted at appropriate places in the quality assurance system to determine whether it is appropriate to advance to the next phase of new product development.

---

* The material in this section was taken from Kurogane, *Cross-Functional Management,* pp. 85–97. It was rewritten to integrate it with the writing style used in this book. The author takes sole responsibility for any errors due to rewriting.

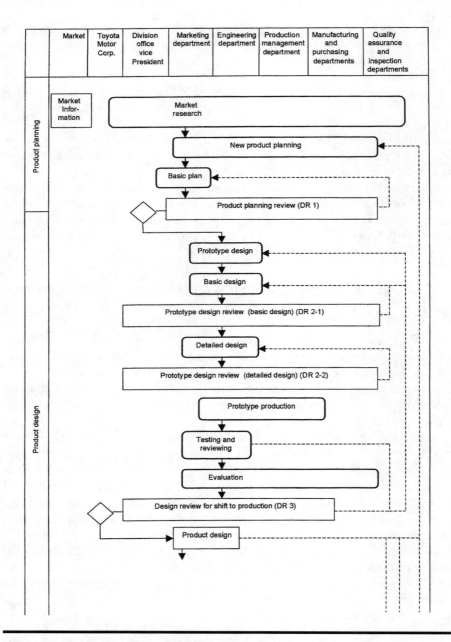

Figure 5.2 Toyota's Cross-Functional Quality Assurance Process

Reprinted from *Cross-Functional Management: Principles and Practical Applications*, Kenji Kurogane, Ed. © 1993 by the Asian Productivity Organization, Tokyo. Reprinted by permission of the Asian Productivity Organization.

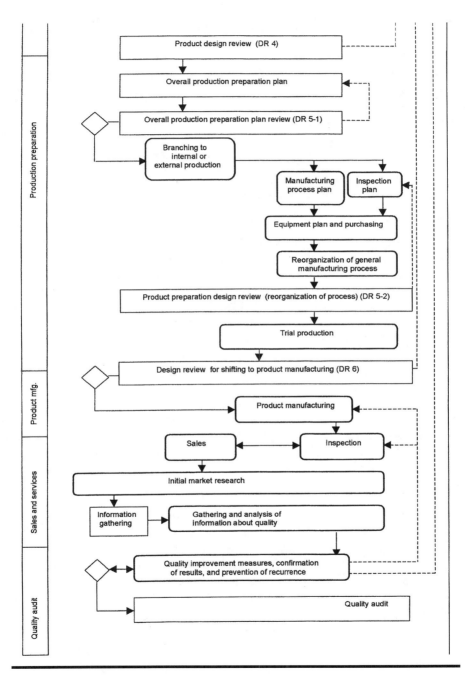

Figure 5.2 (continued) Toyota's Cross-Functional Quality Assurance Process

Quality Assurance Activity in the Development of the X300 Forklift

Overview

The Toyota design review system was used in the development of the X300 forklift at Toyota Forklift. This system integrated six design reviews into the development of the X300 forklift.

The basic idea behind the development of the X300 forklift was to provide an excellent product through farsighted prediction of market trends and customer needs, and to win customer satisfaction and trust.

Quality Assurance in New Product Development

Quality Improvement in Product Planning

The product planning system used to develop the Toyota X300 forklift consists of three phases: market research, new product planning, and developing and reviewing the product plan (Figure 5.3).

The market research phase involves getting a grasp of customer needs and wants by market segment (through surveys) and translating those needs and wants into "demanded quality characteristics."

The new product planning phase involves studying competing products, establishing which "demanded quality characteristics" will stimulate customers to buy in each market segment (called sales points), reviewing the time to market (using Gantt charts, PERT/CPM, or other scheduling methods), reviewing the production costs of the product, and forecasting the demand for the product in each market segment.

The "develop and review" plan phase involves drafting a development plan for the X300 forklift and conducting a product planning review.

Quality Improvement in Product Design

The product design system used to develop the X300 forklift consists of six phases: prototype design, prototype production, test and review, evaluation, shift to production, and product design (Figure 5.4).

The prototype design phase involves conducting an "engineering policy review" for the new product functions. This includes developing a detailed list of relevant processes, parts, mechanisms, and functions with specifications, preparing a critical functions evaluation report, and performing bottle-neck engineering of relevant processes.

The prototype production phase involves a detailed design review of the X300 forklift.

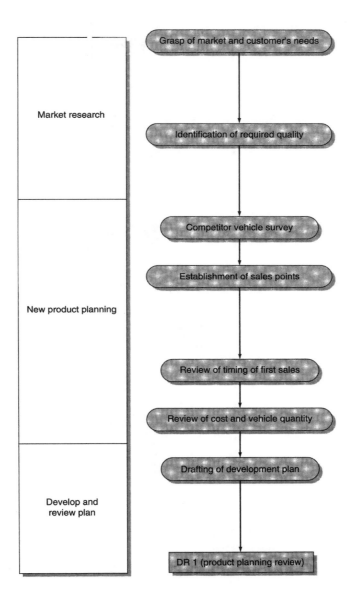

Figure 5.3 Toyota's Product Planning System

Reprinted from *Cross-Functional Management: Principles and Practical Applications*, Kenji Kurogane, Ed. Copyright © 1993 by the Asian Productivity Organization, Tokyo. Reprinted by permission of the Asian Productivity Organization.

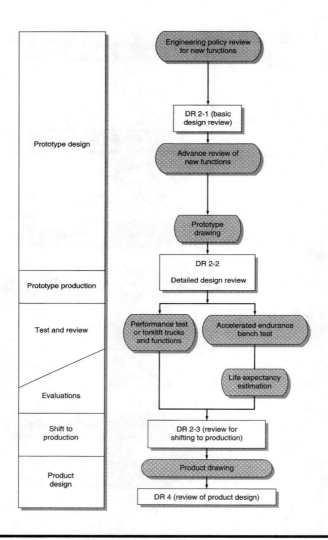

Figure 5.4 Toyota's Product Design System

Reprinted from *Cross-Functional Management: Principles and Practical Applications*, Kenji Kurogane, Ed. Copyright © 1993 by the Asian Productivity Organization, Tokyo. Reprinted by permission of the Asian Productivity Organization.

The test and review and evaluation phases involve establishing test conditions and evaluation criteria through surveys of actual usage conditions and an accelerated endurance bench test. Life expectancy was estimated on the basis of test results and survey data. The above activities increased the degree of comfort at Toyota Forklift in predicting that the design of the X300 was going according to plan and would require few modifications later in its life cycle.

The shift to production phase involves a pass or fail review to shift to trial production.

The product design phase involves the finalization of detailed product drawings and a product design review to determine conformance of design quality to overall quality specifications.

Quality Improvement in Production Preparation

The production preparation system used to develop the X300 forklift consists of eight phases: developing a general production plan, developing a manufacturing process plan, developing an equipment plan, purchasing equipment, reorganizing individual processes to ensure machine capability, reorganizing the entire production process to ensure system capability, trial production, and shifting to product manufacturing (Figure 5.5).

The general production plan involves obtaining confirmation of existing product and process problems, conducting a review to determine if those problems have been resolved, and identifying the characteristics of machines and equipment that were deployed in the product design phase.

The manufacturing process plan involves conducting failure modes and effect analysis (FMEA) on the characteristics of machines and equipment, and discovering any bottleneck problems resulting from equipment and production methods.

The equipment plan phase involves determining the specifications of the equipment required to manufacture the X300 forklift for cost estimation purposes.

The equipment purchasing phase involves confirming the capability of machines and equipment after specifications have been costed out, generating purchase orders for machinery and equipment, and conducting a design review for the completed machines and equipment processes.

The individual process reorganization phase involves ensuring the capacity of individual machines, and the overall process reorganization phase involves ensuring the capacity of the entire production process. These phases include establishing work procedures, developing operating standards, allocating human resources, developing training programs, and surveying appropriate people to predict the capabilities of machines and the entire system.

The trial production phase and the shift to manufacturing phase involve a pass or fail review of the decision to shift the X300 forklift to manufacturing.

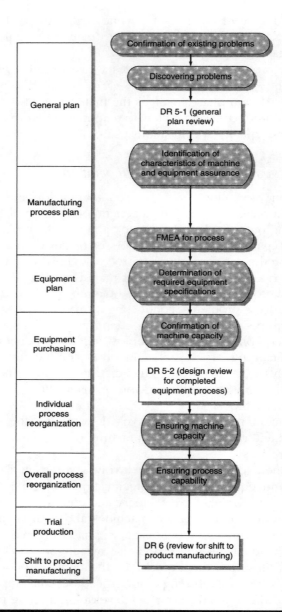

Figure 5.5 Toyota's Production Preparation System

Reprinted from *Cross-Functional Management: Principles and Practical Applications*, Kenji Kurogane, Ed. Copyright © 1993 by the Asian Productivity Organization, Tokyo. Reprinted by permission of the Asian Productivity Organization.

Concluding Remarks

From the inception of quality management activity at Toyota, top management promoted cross-functional management across divisions to upgrade company-wide systems. As a result, the company successfully solved many problems, promoted standardization of systems, and achieved efficient management. The X300 is a case study in how a cross-functional process, developed using cross-functional management, was used for new product design.

A SERVICE APPLICATION: FIELD OF FLOWERS

Background

Field of Flowers is a retailer of flowers and related items located in Davie, Florida. Its president and top management studied Dr. Deming's theory of management and set up the company according to those principles.

At the time the company was first organized, a performance appraisal system had to be developed. Since the leadership of Field of Flowers wanted it to be in keeping with the System of Profound Knowledge, a cross-functional team composed of the top management of selected departments was formed. They used the five step cross-functional management model discussed earlier in this chapter.

First, the team stated a mission for the proposed performance appraisal system: "To develop a performance appraisal system consistent with the System of Profound Knowledge." They did this and made it clear to all employees.

Second, team members identified all stakeholders of the performance appraisal process: candidates, new hires, employees, supervisors, and top management.

Third, the team constructed an integrated flowchart of a traditional human resource system, with special attention to the performance appraisal functions; see the shaded sections in column one of Figure 5.1.

Fourth, team members identified key measures of the efficiency and effectiveness of the performance appraisal system. The efficiency of the performance appraisal process is measured by the percentage of performance appraisals completed on time, by supervisor and overall, by year. The effectiveness of the performance appraisal process is measured by the following key indicators:

1. The percentage of performance appraisals with written comments offering ideas for improvement of work
2. An analysis of responses from employees to the questions, "Did you know what to do to improve your job performance upon leaving your performance review?" "Did your supervisor?" by year.

Fifth, team members developed modifications to the integrated flow-chart shown in Figure 5.1 based on the System of Profound Knowledge. The ideas for the modifications came from the work of Peter Scholtes.* Scholtes identified the following functions as the components of a performance appraisal system: (1) provide feedback to employees on their work; (2) provide a basis for salary increases and bonuses; (3) identify candidates for promotion; (4) provide periodic direction of an employee's work; (5) provide an opportunity to give recognition, direction, and feedback regarding special projects; (6) identify needs for training, education, and skill or career development; (7) provide an equitable, objective, defensible system that satisfies the requirements of the Civil Rights Act and the Equal Opportunity Commission guidelines; and (8) provide a channel for communication.

The Revised System

The *provide feedback for improvement* step in the *Performance Evaluation* section of Figure 5.1 is redefined to be providing an employee feedback on his or her work. Feedback can be provided by identifying the major processes in which the employee is involved, identifying the major work group(s) to which the employee belongs, developing a list of major feedback resources for the employee (e.g., key customers and suppliers), and developing an agenda and method for obtaining feedback from each feedback resource.

The *wage and salary determination, raises, bonuses, and other monetary issues* steps in the *Compensation Management* and the *workers' compensation* step in the *Benefits Management* section of Figure 5.1 are redefined to be providing a basis for salary increases and bonuses based on market rate (what it would cost to replace someone on the open market), accumulation of skills (flexibility due to acquired abilities), accumulation of responsibility (depth of contribution to a greater number of processes and influence over a larger number of employees), seniority within an organization and within a job classification, and prosperity (profit-sharing of the entire organization, not one segment of the organization).

The *change in employee's position [promotion or demotion]* step in the *Transfers* section of Figure 5.1 is redefined to be identifying candidates for promotion by providing special assignments that contain elements of

---

* The structure of this chapter is heavily drawn from Scholtes, P., "An Elaboration on Deming's Teachings on Performance Appraisal," Joiner Associates, Madison, WI, 1987. The author has modified some of the ideas in Mr. Scholtes' paper and takes sole responsibility for his modifications.

the promotion job, utilizing an assessment center to observe candidates exercising the skills needed in the promotion job under realistic conditions (if available), determining the needs and wants of the stakeholders of the promotion job in respect to the characteristics of the person who will assume the promotion job, and developing an organizational culture in which promotion is not the only vehicle for people to exercise leadership and influence, to get rewards and recognition, or to stretch and challenge themselves in their jobs and careers.

The *familiarize employees with objectives* and *familiarize employee with work expectations* steps of the *Orientation* section of Figure 5.1 is redefined to be providing periodic direction of employees by communicating the organization's strategic and business plans to help each employee define his or her work, and spending time with each employee to develop methods to promote the organization's strategic and business plans.

The *job enhancement* step of the *Development (Managerial Skills)* section and the *appraise subordinate's behavior* step of the *Performance Evaluation* section of Figure 5.1 are redefined to be providing an opportunity to give recognition, direction, and feedback to an employee regarding his or her work on special projects.

All of the training steps in the *Training (Vocational Skills)* section of Figure 5.1 are redefined to be identifying each employee's needs for training through the empowerment process; that is, each employee receives the training required to turn the SDSA and PDSA cycles for process standardization and improvement.

The *forecast employment requirements* step of the *Human Resources Planning* section, the *locate qualified candidates* step of the *Recruitment* section, the *select hire* step of the *Selection* section, and the *fired* and *layoff* steps of the *Terminations* section of Figure 5.1 are redefined to be providing an equitable, objective, defensible system that satisfies the requirement of the 1964 Civil Rights Act and the Equal Opportunity Commission guidelines of 1966 and 1970 can be accomplished by committing to the values and spirit inscribed in the law, not just by conforming to the law.

The *resolve personal problems* and *improve employee performance* steps of the *Employee Relations* section of Figure 5.1 are redefined to be providing a channel for communication that otherwise would probably not occur. This can be accomplished by all employees in an organization asking and answering the following questions: "With whom is it important to maintain communication? For what purpose? With what frequency? In what kind of setting, format, or agenda?" Answers to the above questions should promote the flow of information and knowledge into channels of communication between people in organizations.

It is important to realize that all of the above processes form an interdependent system of processes. It does not make sense to adopt a new process for providing employees a basis for salary and bonuses and not provide them a process for identifying needs for training, education, and skill or career development. To do so may create a monster worse than the existing system of performance appraisal. For example, guaranteeing salary based on seniority without any process to improve the employee or organizational processes could be a formula for disaster.

Given the integrated flowchart in Figure 5.1 and redefinition of the above steps in the flowchart, the Field of Flowers cross-functional team reconceptualized the performance appraisal process. The results are presented in the following excerpts, paraphrases, and unwritten understandings from the Field of Flowers Employees Handbook.

Excerpts are noted in regular typeface. Unwritten understandings are in italics. Comments by the author are noted in Serif BT type.

Issue 1

Provide employees feedback on his or her work.

> All associates have the following rights: ... the right to have access to all useful, pertinent information about the enterprise and about one's particular job.... This includes the right to participate in the process or decision making related to one's work area.

> If you have questions or concerns about anything related to your employment, talk with your Team Leader. That person will assist you in every way possible.

Issue 2

Provide a basis for salary increases and bonuses.

> *All employees begin their careers with Field of Flowers at the same hourly salary. This applies to everyone: floral arrangers, delivery personnel, sales associates, and so on. The orientation period for every potential associate is based on the needs of that particular individual. Thus, the length of orientation varies from person to person. After orientation is complete, employees become level two associates and their hourly salary is automatically raised to a standard level. After 2 years, level two associates become tenured associates and maintain their hourly salary plus profit-sharing.*

Associates who have achieved tenured status will be beneficiaries of profit-sharing for any year in which profit-sharing is appropriate (according to the established corporate guidelines). Percentage of profit-sharing will be based upon earned income. Part-time and full-time associates will participate in profit-sharing.

All employees know daily sales figures and are provided complete financial disclosure once each year so that they understand the distribution of profit-sharing.

Issue 3

Identify candidates for promotion.

Promotion. We believe in the benefits of advancing people from within the organization whenever possible. We recognize that this requires that numerous very able people be hired into entry level positions and be offered the opportunity to expand their knowledge.

Termination. The following sections of the Employee Handbook deal with termination. We have decided to place these sections under issue 3 because they deal with termination (a form of demotion).

All tenured associates have the right to employment security. In keeping with its philosophy of long-term commitment to its associates, Field of Flowers will consider layoffs of tenured associates only after all other remedies — including reduced profit expectations — have been exhausted. If work force reductions become necessary for the survival of the company, then a council of tenured associates will be formed to advise management as to the best manner in which to make those reductions. In the case of severe disciplinary action or dismissal for cause, tenured associates will have the right to demand peer review of such action. An elected Associates' Council will review such cases; they will have the authority to overturn or revise the action taken.

During the Orientation process, especially during the first 90 days of employment, Associates are expected to be attentive and interested in learning the procedures we follow. After the Trial

Period ends and the Associate is raised to Associate II, they are expected to follow the guidelines and procedures that were a part of their Orientation/Training. Any Associate who deliberately and knowingly refuses to adhere to established procedures can be dismissed without following the usual measures that preclude termination without counseling and documentation.

Issue 4

Provide periodic direction to employee's work.

Field of Flowers' leadership provides feedback to employees by stating and constantly pursuing their mission, taking responsibility for processes, and working with employees to develop plans for the methods needed to achieve the Field of Flowers mission.

Field of Flowers has a statement of mission and values. This statement creates a culture in which leadership provides direction to employees. The statement appears below:

In management, the first concern of the company is the happiness of the people who are connected with it. If the people do not feel happy and cannot be made happy, that company does not deserve to exist.

— Ishikawa

The primary mission of Field of Flowers is to provide stable, safe, fulfilling employment for our family of associates. We realize that the way to accomplish this is to be recognized by the community as the company, in our chosen field of endeavor, which provides the highest quality products and services to its customers.

We believe in the importance of constancy of purpose toward never-ending improvement of the processes which produce our products and services. We further believe that the leadership of the company must take responsibility for these processes. Associates must not be held accountable for improving results if they do not have the authority or the resources to change the processes which produce those results.

We must always be alert to the harm that can result from setting arbitrary numerical goals and standards without providing the methods for achieving them.

Issue 5

Provide an opportunity to give recognition, direction, and feedback to an employee regarding his or her work on special projects.

All associates are given opportunities to grow and develop in ways that are mutually beneficial to themselves and the company. Such growth and development can include special projects. It is the responsibility of management to set into motion and nurture the improvements and innovations developed by associates. This type of leadership will stimulate the intrinsic motivation of associates.

Issue 6

Identify an employee's needs for training, education, and skill or career development.

The Management Team (top management of Field of Flowers) accepts complete responsibility for accurately and adequately training all associates. By empowering associates with knowledge and skill, stress is reduced and their employment experience can be pleasant and rewarding.

We believe in vigorous programs for training and education so that our associates are able to grow as workers as well as in other aspects of their lives.

...training and a supportive attitude on the part of leadership, will empower front line associates to make decisions on their own.

Tuition reimbursement is available to associates with at least 1 year of service. Company approval is required prior to enrollment and will be determined on a case by case basis.

Issue 7

Provide an equitable, objective, defensible system that satisfies the requirement of the 1964 Civil Rights Act and the Equal Opportunity Commission guidelines of 1966 and 1970.

We are committed to selecting the most qualified person for each position in our company. Our success is dependent upon our maintaining high standards and emphasizing "teamwork!" All personnel selections are in accordance with Equal Employment Opportunity guidelines.

Each employee contributes to Field of Flowers' success; each will be treated fairly.

Issue 8

Provide a channel for communication.

The company culture of Field of Flowers promotes open, multi-way communication within and between all levels of employees. The performance appraisal process at Field of Flowers is a daily ongoing process of communication that constantly seeks to increase employees' ability to take pride in their work and joy in the outcome, and to optimize its interdependent system of stakeholders.

The development of the performance appraisal system was the first step (Standardize) of the SDSA cycle. The team progressed through the Do, Study, and Act stages and now continuously works on improvement of the performance appraisal system through application of the PDSA cycle.

In addition to the above eight issues, the management team at Field of Flowers worked continuously to improve their human resource planning, recruitment, selection, and orientation processes. They believed that the need for the remedial aspects of performance appraisal are inversely related to the quality of the people that enter their organization. Conversely, they believed that the need for the constructive aspects of performance appraisal are always present in an organization.

SUMMARY

Chapter 5 discusses cross-functional management, Prong Two of the Quality Management model presented in this book. Cross-functional management is important because it weaves together the vertical (line) functions of management with the horizontal (interdepartmental) functions of management. Primary applications of cross-functional management include quality management, cost management, delivery management, and personnel management. Other applications are new product development, sales management, and safety management.

Selecting cross-functional teams was discussed in this chapter. The members of the EC initially form cross-functional teams and select their leaders. The leader, who is an executive in charge of a function, recommends the members for the team, preferably no more than five people. It is not necessary for all team members to come from affected areas. All team members are trained in appropriate theory and practice.

In small organizations, one cross-functional team comprising all relevant executives can be established to coordinate and optimize all company-wide systems. In large organizations, one cross-functional team can be set up for each company-wide system, such as quality management, safety management, or personnel management. The EC reviews, manages, and coordinates all cross-functional teams. Cross-functional management reviews are conducted at least yearly by the cross-functional team leader.

Implementing cross-functional management is difficult because of its interdisciplinary nature. To ensure the success of a cross-functional team, it is created with the expectation that it will be permanent and deal with continuous improvement of a company-wide system over the long term. Cross-functional team members learn to think in terms of the whole system, not just their areas. Communicating the results of the cross-functional team's work is extremely important.

Chapter 5 presents a generic cross-functional management system for cutting costs in a standardized fashion across the different areas in an organization. Also, it presents a cross-functional management system for new product development used in a manufacturing company, Toyota Forklift. Finally, it discusses how a service company, Field of Flowers, used a cross-functional team to create its performance appraisal system.

6

PRONG 3:
POLICY MANAGEMENT

PURPOSE OF THIS CHAPTER

The purpose of this chapter is to explain what is required to set policy, deploy policy, implement policy, study policy, provide feedback to employees on policy, and conduct presidential review of policy in an organization. Policy management is Prong Three of the Quality Management model presented in this book.

BACKGROUND

Policy management* is performed by turning the PDSA cycle to improve and innovate the methods responsible for the difference between corporate results and corporate targets or to change the direction of an organization. Corporate targets are set to allocate resources between corporate methods. Policy management assumes that daily management and cross-functional management are at work in the organization.

---

* The material on policy management is drawn heavily from the following sources: (1) Mizuno, S., *Management for Quality Improvement: The 7 New QC Tools*, Productivity Press (Cambridge, MA), 1988. (2) Ishikawa, K., *What is Total Quality Control? The Japanese Way*, Prentice-Hall (Englewood Cliffs, NJ) 1985, pp. 59–71. (3) King, B*., Hoshin Planning: The Developmental Approach*, GOAL/QPC (Metheun, MA), 1989. (4) Brunetti, W., *Achieving Total Quality: Integrating Business Strategy and Customer Needs*, Quality Resources (White Plains, NY), 1993. The author would like to thank J. Michael Adams and Francisco "Tony" Avello of Florida Power & Light Co. for their input. The author takes sole responsibility for the material in this section.

Policy management is accomplished through an interlocking system of committees (see Figure 6.1). The Executive Committee (EC) is responsible for setting the strategic plan for the entire organization. That includes establishing values and beliefs, developing statements of vision and mission, and preparing a draft set of strategic objectives. The Policy Deployment Committee (PDC) is responsible for deploying the strategic objectives in the entire organization. That includes developing an improvement plan (set of short-term tactics) for each department. A Local Steering Team (LST) is responsible for implementing policy (short-term tactics) within a department by coordinating and managing project teams. Project teams implement policy through improvement and innovation of the processes highlighted for attention.

The Local Steering Teams conduct meetings with Project Teams, called Feedback and Review sessions, to learn about team activity, promote quality theory and tools, and manage and coordinate team activities to

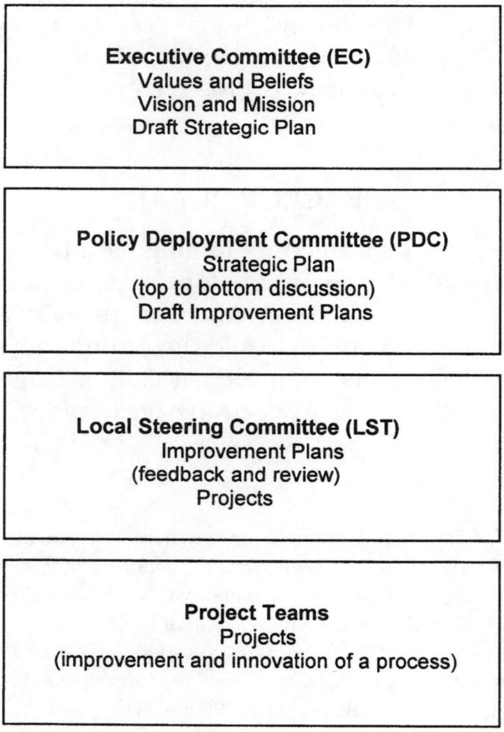

```
┌──────────────────────────────────────┐
│      Executive Committee (EC)          │
│        Values and Beliefs              │
│        Vision and Mission              │
│        Draft Strategic Plan            │
└──────────────────────────────────────┘

┌──────────────────────────────────────┐
│  Policy Deployment Committee (PDC)     │
│           Strategic Plan               │
│       (top to bottom discussion)       │
│       Draft Improvement Plans          │
└──────────────────────────────────────┘

┌──────────────────────────────────────┐
│   Local Steering Committee (LST)       │
│         Improvement Plans              │
│        (feedback and review)           │
│             Projects                   │
└──────────────────────────────────────┘

┌──────────────────────────────────────┐
│          Project Teams                 │
│             Projects                   │
│ (improvement and innovation of a process)│
└──────────────────────────────────────┘
```

Figure 6.1 Committee Structure for Policy Management

pursue company policy. The Policy Deployment Committee conducts meetings with Local Steering Committees, called Mini-SITCONS, to learn about team activity, promote quality theory and tools, coordinate and manage project teams to optimize company policy, and, if necessary, to reallocate resources between project teams (according to revised targets). Finally, the President meets with the leader of each department to understand the state of quality in the organization and to determine if policy (strategic objectives) is being implemented throughout the organization.

INITIAL PRESIDENTIAL REVIEW

The President conducts an initial Presidential Review (see step 28 of the Detailed Fork Model in Figure 1.2) to determine the state of the organization and to develop a plan of action for the promotion of corporate policy. Presidential Reviews are high-level studies of an organization's departments by the President or Chief Executive Officer.*

During Presidential Reviews, the leaders of the departments explain to the President their mission and the status of projects emanating from the strategic and improvement plans. Normally, this information is conveyed through presentations. Much attention is devoted to the linkage between corporate and department strategies, and the progress toward the achievement of these strategies. Problems in planning and executing these strategies are discussed, and attempts are made to identify the causes of these problems. Through the Presidential Review, the President is able to evaluate the state of quality and management in the organization.

Reasons for Conducting the Presidential Review

Presidential Reviews are conducted for several reasons. First, they are conducted to determine the extent of achievement of organizational policy. Reviews are conducted to verify the implementation of improvement plans and to assess and improve the management process used to achieve the mission. In one company, the President found out that one of his policies had been completely misinterpreted, and the troops were marching in the opposite direction. The mistake was identified and quickly rectified to avert much wasted effort. This is not a rare occurrence in large organizations because information is filtered by each layer of management. Second, Presidential Reviews are conducted to determine the cost to the organization of achieving its strategic and improvement plans. Third, Presidential Reviews

---

* This section of the chapter was rewritten from material prepared by Francisco "Tony" Avello of Florida Power & Light Co., Miami, Florida, 1992.

are conducted to prevent deterioration in methods not highlighted for attention in policy management, due to the reallocation of resources to methods highlighted for attention in policy management. Finally, Presidential Reviews identify the major problems facing the organization.

The President tries to discover those problems that affect functional performance but cannot be solved at the functional level. Generally, these problems have to be addressed at the company level, since the causes cross many organizational boundaries. In this way, no single function has the authority to promote solutions. Most major company problems are cross-functional and thus difficult to identify. Because of its cross-functional nature, the Presidential Review provides a significant opportunity to identify these problems. Once identified, these problems are turned over to appropriate cross-functional teams.

Benefits of Presidential Reviews

One benefit of Presidential Reviews is that they create a dialogue between the President and mid-level management. This dialogue encourages an atmosphere of trust that helps bring out information about problems. The information provides an opportunity for the President to promote joy in work and pride in the outcome for all employees.

Another benefit of Presidential Reviews is the insight they give to the President about the operations and culture of the organization. Frequently, this information is not available through normal channels of communication. Examples of information that can be gleaned by the President include the skill level of the managers and supervisors, the attitudes of employees toward improvement of methods, and employee morale. This information is necessary to promote the strategic and improvement plans.

The President will have a good understanding of the major problems facing the organization after a full round of Presidential Reviews. So, to a certain extent, he or she should have a good idea about the possible causes of problems. The President knows the areas that should be involved in the improvement activities. He or she should also know the attitudes and skills of employees in carrying out the strategic and improvement plans. Finally, he or she knows the level of training that will be needed throughout the organization to work on the strategic and improvement plans.

Barriers to the Presidential Review

Initially, the President may resist conducting Presidential Reviews due to demands on his or her time. All too often, there is a desire to obtain information from an executive summary. However, the executive summary does not provide sufficient information to establish or change the direction of the company. One company President tells the story of how he went from opposing Presidential Reviews to so thoroughly embracing them that he began to conduct half-day reviews on a quarterly basis with each of his departments.

Selecting the Departments and Topics to Review

Departments and topics are selected for Presidential Review by examining the policies and projects that were not successful in previous years. Underachieved policies and projects are considered failures of the management system. These problems identify the departments that are candidates for Presidential Review. It is important that the President doesn't assign fault for problems. Blame-fixing makes people defensive and unwilling to identify problems. It creates fear in the work place. The President takes responsibility for problems in the system.

Another way of selecting departments and/or topics for Presidential Review is to proceed as explained in the previous paragraph, but to review all departments. This has the advantage of not singling out any department's past failures, thus avoiding a threatening situation. One drawback to this approach is that more departments have to be reviewed than in the first alternative. Ultimately, the culture of the company and the existing organizational climate will dictate which alternative is best. In either scenario, the issue of not creating a threatening situation is an important one and is weighed carefully before deciding which approach to take.

Informing the Departments to Be Reviewed

Once the topics and functions have been identified, the next step is to announce the reviews. This is done through a meeting of senior managers. The purpose of the reviews is explained in this meeting. The names of the departments that will participate in the reviews are announced, and the format of the reviews is discussed. Steps are taken to put the participating departments at ease. If needed, the President offers staff help in further clarifying the objectives, guidelines, and manner of the reviews. Also, this is a good time to define the ground rules to follow during the reviews.

Ground Rules for the Presidential Review

Probably the most important ground rule for Presidential Review is that the presenter submits his or her department's report at least 1 week prior to the review. This rule is usually resisted, since most presenters will make changes to their presentation until the last minute. However, as will become apparent in the next section, it is important to enforce this rule. Another important ground rule is using data to support the points of the presentation. Since the reviewer will be using the presentation as a vehicle to acquire information for establishing company policies, the presentation should rely on facts as much as possible.

The presenting department is allowed to bring and use as many presenters as needed to fully explain the principal issues or to answer questions. The President usually invites managers from related departments to the review. This is done not only to make them aware of the important issues of that department, but also for them to get a glimpse of the review procedures and thus help them prepare for their own reviews. The atmosphere of the review is informal but serious.

Preparing for the Reviews

Proper preparation for a Presidential Review is important. Many reviews fail before they begin because of poor preparation by the President. Good reviews are the result of careful study of the presenter's report before the reviews. This study allows the President to establish a focus for the review, identify issues needing clarification, and formulate questions.

It is critical to have a staff department help the President prepare for the reviews. Usually, this task is assigned to the Quality Department. However, it could be any department knowledgeable about the Presidential Review process and Quality Management in general. The assigned department assists the President in fully understanding the present situation of the presenting department and in developing a list of topics or broad questions to ask the presenters. More specific questions will normally follow from the answers given by the presenters. The President conducts the review and becomes knowledgeable enough to conduct future reviews without extensive help. Therefore, a key task of the staff department assisting the President is to instruct and coach the President so that he or she can become a competent reviewer.

Conducting the Review

Usually, the review begins with a presentation by the management of a department, not by the team members who performed the work in the department. This structure is critical because it places management "in the line of fire" for quality improvement efforts. Management must be involved with, and in learning about, quality management to effectively perform their responsibilities in a presidential review. The presentation is followed by a question and answer period that is led by the President. It is customary to allow the presenter to finish the presentation without interruption, except for clarifying questions.

When the presenter has concluded, the President begins the question and answer period. It is his or her opportunity to probe deeply into issues to determine the possible causes of problems. Often, the President will be persistent and ask the same question several times to get the appropriate answer. For critical issues or when the answers are not provided, action items with due dates are established. The presenters demonstrate the results of the action items at a later date.

The following questions are examples of the type of questions the President might ask a presenter. They are offered here for illustration purposes only.

1. What is the mission of your department?
2. Does your department's mission support the company's mission?
3. How do you know if you are pursuing company policy?
4. What procedures do you follow when you discover that you are not pursuing company policy?
5. Can you show me an example of a corrective action you have taken when your department was not pursuing company policy?
6. How did you analyze the failure to pursue company policy?
7. How did you know if the corrective action was effective?
8. What are the major problems/opportunities of your department?
9. How do these problems/opportunities manifest themselves?
10. Can you give me an example?
11. What effects do these problems/opportunities have on your department and/or on your customers?
12. Who are your customers?
13. What are your customers' needs and wants?
14. How, and how often, do you assess your customers' needs and wants?
15. Can you show me how you are ensuring the satisfaction of your customers' needs and wants?

The attitude of the President during the review is very important. Often, the President has to be persistent to obtain the answers he or she needs to make decisions. In some cases, the President pushes the presenters to obtain a desired performance level or behavior. This may be seen as judgmental or harsh by the presenters. The President's job is to establish an atmosphere of teamwork, providing constructive criticism, examples of ideas for improvement, and/or where to go for help.

The staff department members and consultant helping the President during the review assume a low profile. They ask questions only at the request of the President, after the President has finished his or her own questions.

After the reviews, the staff department and/or the consultant meets with the President to identify his or her successful and unsuccessful actions and behaviors during the review. The purpose of this meeting is to instruct and coach the President to improve his or her skills as a reviewer. It is best to concentrate on only two to three items at one time and not tamper with the Presidential Review process.

Keys to Successful Reviews

The most important determinant to a successful Presidential Review is whether the President can gain the trust of management. It is critical that the President guarantees that presenters will not be penalized if they disclose problems in their departments.

Another key to a successful review is the quality of the preparation by the President and his staff. If they learn all they can about the topic being reviewed, research and analyze the past accomplishments and failures of the department being reviewed, and focus on problem areas that offer good opportunities for improvement, they are more likely to create a positive review process.

Another important factor in a successful review is the assignment of action items to presenters when data is not provided as requested. Failure to assign action items when needed may communicate to the organization that mediocrity is acceptable.

It is more important to concentrate on process and results, not only results, in Presidential Reviews. The President helps the presenter see how poor results are most likely due to a deficiency in a management process. Also, the President sets an example by identifying and working to improve deficiencies in the Presidential Review process.

Presidential Reviews and Daily Management

The early phases of daily management should include the Initial Presidential Review process (see step 28 of the Detailed Fork Model in Figure 1.2). This is important to ensure the active involvement of all levels of management in the Quality Management system.

POLICY SETTING

Once the initial Presidential Review is complete, the President has information critical to setting policy. Policy setting involves (1) establishing statements of vision and mission, (2) developing organizational values and beliefs, (3) identifying organizational and environmental factors that effect policy, (4) identifying crises facing the organization, (5) determining key organizational processes that effect stakeholders, (6) identifying technological issues facing the organization, (7) establishing strategic objectives (see step 29 of the Detailed Fork Model in Figure 1.2), and (8) developing set of integrated improvement plans (see step 30 of the Detailed Fork Model in Figure 1.2) for the departments.

Executive Committee

The members of the Executive Committee (EC) work to understand the pros and cons for transformation of the organization from the perspective of each stakeholder group. The top management of an organization asks the question: "Does my organization have the motivation and energy necessary to make quality happen?" The data from this question is collected, summarized, and analyzed through a force field analysis. Force field analysis is a technique which lists the "forces for" and "forces against" a particular action or issue.

After analysis, if the members of the EC determine that the "forces for" transformation outweigh the "forces against" transformation, they develop a strategic plan for transformation. The comparative weight of the "forces for" transformation versus the "forces against" transformation is a subjective decision on the part of top management.

A strategic plan lists the long-term strategic objectives of an organization. Strategic objectives are based on a thorough analysis of statements of vision and mission, values and beliefs, organizational and environmental factors, crises, if any, key processes that affect stakeholders, and technology. Figure 6.2 depicts the relationship between the above six items and the strategic objectives.

Statements of Vision and Mission

Dr. Deming's theory of management addresses the need to establish constancy of purpose toward improvement of product and service with a plan to become competitive, stay in business, and provide jobs (see point 1 of Dr. Deming's 14 points). Statements of vision and mission are starting points for constancy of purpose.

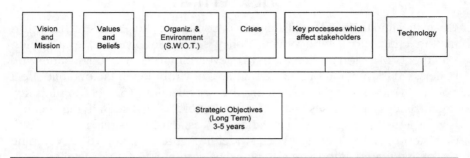

Figure 6.2 Development of Strategic Objectives

A vision statement is developed by the top management of an organization. It defines the organization's future state. It is a dream that comes from the "hearts" of top management. It should evoke emotion, be easily remembered, state a noble purpose, and create a rallying point for all concerned with the organization.

The mission statement reveals the current reason for the existence of an organization. Top management develops it. The mission statement provides a rallying point for all stakeholders. It should be short and easily remembered by all stakeholders of the organization.

Values and Beliefs

A statement of the values and beliefs that govern an organization's culture is necessary to provide predictable uniformity and dependability to the decision-making process. This statement forms the foundation for the decision-making process. Values and beliefs are theories about life and organizations that have been modified and improved by cultural, educational, familial, organizational, and personal experiences. However, through the adoption of Dr. Deming's theory of management, an organization can develop a set of values and beliefs that form its foundation. The values and beliefs inherent in Dr. Deming's theory of management focus on creating a "win–win" environment for all stakeholders of an organization. They are discussed below.

- Manage to optimize the entire system, not just your component of the system. Local optimization creates suboptimization of the entire system. For example, maximizing profit in one division of a company may decrease profit for the entire company.
- Manage to create a balance of intrinsic and extrinsic motivation; do not just motivate people using extrinsic motivation. Intrinsic

motivation is the individual's desire to do something for its own value, as opposed to extrinsic motivation that relies on rewards or punishments for the individual. For example, empower people to promote joy in work, as well as using well-thought-out pay plans.

■ Manage with a long-term process and results orientation, not with a short-term results-only orientation. Process and results management promotes improvement and innovation of organizational processes. Highly capable processes facilitate prediction of the future, and consequently, a higher likelihood of achieving the organizational mission. For example, study a process, collect data, and develop ideas for improving the process to permanently remove problems, as opposed to just demanding fewer problems.

■ Manage to promote cooperation, not competition. In a competitive environment, most people lose. The costs resulting from competition are unknown and unknowable, but they are huge. Competition causes individuals, or departments, to optimize their own efforts at the expense of other stakeholders. This form of optimization seriously erodes the performance of the system of interdependent stakeholders. For example, get departments and divisions to work together for the common good, as opposed to working against each other for their individual good.

Organizational and Environmental Factors

A S.W.O.T. (Organizational Strengths and Weaknesses, and Environmental Opportunities and Threats) analysis is used to assist the members of the EC in selecting the strategic objectives that ensure the best fit between the internal strengths and weaknesses of an organization and the external opportunities and threats that face an organization. The members of the EC identify strengths and opportunities that bypass weaknesses and threats. This information is then used as input in developing the strategic objectives of the organization.

An excellent method for performing a S.W.O.T. analysis is for the members of the EC to appoint a team of "appropriate" people to conduct four brainstorming sessions and to create four affinity diagrams. One brainstorming session is conducted on the organization's strengths and an affinity diagram is developed to bring out the underlying structure of the strengths. Similar analyses are conducted for weaknesses, opportunities, and threats.

Portions of an affinity diagram from a S.W.O.T. analysis of a university are shown in Figure 6.3. The items on the left side of Figure 6.3 provide

Strengths

Academic   (faculty, research, library, …)
Financial   (management system, …)
Diversity   (multicultural, …)
Business   (continuousimprovementprocess,abilitytoplan,training)
Life Style   (encourages holistic health, total fitness centers, …)

Weaknesses

Communication   (ineffective, …)
Infrastructure (space, parking, funding, …)
Financial   (tuition, budget, …)
Reputation   (local v. international sports, …)
Technology   (insufficient use, …)

Opportunities

Student Base   (local, international, retrain downsizers, …)
Resources   (donations, local economy, alliances, media, …)
International   (sister universities world wide, globalization)
Technology   (easy access to potential students, …)

Threats

Increased Costs   (technology, salaries, tax charges, insurance, …)
Corporate Support (employee tuition reimbursement, downsizing, …)
Litigation   (litigious society, …)

Figure 6.3 S.W.O.T. Analysis for a University

the structure of the strengths, weaknesses, opportunities, and threats, while the items in parenthesis on the right side of Figure 6.3 provide the brainstormed ideas underlying the structure of the strengths, weaknesses, opportunities, and threats.

Crises

The members of the EC determine if any crises currently face the organization. If crises are currently facing the organization, the members of the EC communicate this information to all stakeholders to create the energy necessary to improve quality. Management uses the output from the brainstorming session and affinity diagram in step 1 of the Detailed Fork Model in Figure 1.2 (see Chapter 2) to highlight existing crises. Highlighting crises in this fashion is an important job of top management. Again, it is necessary for leadership to isolate a crisis to generate the energy necessary to improve quality.

Key Processes That Affect Stakeholders

Data are collected to determine the requirements important to the customers served by an organization (called "Voice of the Customer," see Appendix 6A) and the requirements important to all levels of employees in the organization (called "Voice of the Business," see Appendix 6B). Determination of these issues will help the members of the EC identify the key processes (methods) whose level of performance will affect the selection of strategic objectives.

"Voice of the Customer" and "Voice of the Business" data are summarized into a single prioritized list of the organizational processes to be highlighted for attention through the strategic objectives. The tool used to prioritize customer and employee issues is called the Table of Tables (see Appendix 6C). There are many possible structures and scoring schemes for a Table of Tables; see Figure C.1 of Appendix 6C for one possible structure.

Table 6.1 shows another possible structure and scoring scheme for a Table of Tables; it shows sections of a Table of Tables for a university. The university Table of Tables has stakeholder groups and their needs/wants in the rows and university processes in columns 2 through 16. Column 1 shows the average total weight of each stakeholder need/want from a survey conducted in each stakeholder segment. Total weight is a measure of importance and severity to stakeholders. It ranges from 1 to 25, where 1 is not important and not severe and 25 is important and severe. The numbers in the cells indicate the strength of the relationship between a stakeholder need/want and a university process. Relationship values are measured on a 0 (blank) to 9 scale, where 0 (blank) indicates no relationship and 9 indicates a strong relationship. For each cell in a given row, the average total weight for the row is multiplied by the cell relationship value to yield a cell value. Finally, the cell values are summed up for each column. This assumes that all stakeholder groups are equally important when developing strategic objectives. The column totals provide a weighted value for each university process in satisfying important and severe stakeholder needs/wants. In this example, "Employee Training & Hiring" (weight = 5244) and "Communication Systems" (weight = 4066) are the university processes most critical to satisfying stakeholder needs and wants. These are the inputs from the Table of Tables to establishing strategic objectives for the university.

Technology

The members of the EC collect data on technological advances of products and services in the industry, substitute products and services, future

Table 6.1 Portions of the Table of Tables for a University

Stakeholder Groups	Total Weight (1)	Employee Hiring & Training (2)	Communication Systems (3)	Strategic Planning (4)	Facilities Administration (5)	Building & Expansion (6)	Student Training (7)	... (8)	... (9)	... (10)	... (11)	... (12)	... (13)	Grant Administration (14)	Food Services (15)	... (16)
Students																
Value of education	18	9	9	9	1	3	9							0	0	
Quality of academic major	18	9	6	3	0	3	9							0	0	
Quality of instruction	16	9	6	3	3	3	9							0	0	
Availability of courses	14	3	3	1	0	1	0							0	0	
Accuracy of financial awards	16	1	1	3	0	9	0							0	0	
Career preparation	14	9	9	3	0	1	9							0	0	
Attitude of faculty	16	9	3	3	0	0	3							0	0	
Academic reputation	16	9	9	9	3	1	9							0	0	
Quality of academic advising	14	9	3	3	0	3	1							0	0	
Challenge offered by program of study	16	9	3	3	0	3	2							0	0	
Relationship with other students	9	3	3	3	3	0	2							0	0	
Class size	9	1	3	3	1	1	1							0	0	
Buildings and Grounds																

University Processes

...

	Weight	Granting Agencies	Employers	Administration	Staff	Faculty	Board of Directors	Donors	State Government	Federal Government	Etc.	Process Weights
Design new facilities to minimize life cycle needs	20	1	6	9	9	9	9	9	0	0	8	3
Consolidate authority for buildings	9	0	3	1	9	9	3	0	0	0	6	8
Schedule classrooms to minimize costs	8	1	0	1	9	9	9	1	0	0	5	6
⋮												
Process Weights	.5	5	4	3	2	2	2	2	2	4	4	.5

products and services, and management technology. All forms of technology are considered when establishing policy via strategic objectives. The same procedure is used to analyze technology data as is used to analyze crisis and S.W.O.T. data.

Develop Strategic Objectives and a Budget

The members of the EC utilize the information gathered in the above six areas to create a short list of three to five strategic objectives on which the organization will focus extra effort in the next 3 to 5 years, through policy management. Next, the members of the EC establish an initial budget to allocate resources between strategic objectives. It is important that strategic plans identify all available resources (for example, financial, human, and plant and equipment). Resources are allocated to methods by setting targets. Targets may be reset at a later date to optimize the interdependent system of stakeholders of an organization. Strategic objectives are things that will be done in addition to the regular functioning of the organization.

Policy Deployment Committee

The members of the EC communicate to the members of the Policy Deployment Committee (PDC) the pros and cons for transformation, vision and mission statements, values and beliefs, and strategic objectives.

The members of the PDC develop a set of integrated improvement plans (step 30 in the Detailed Fork Model in Figure 1.2) to promote the strategic objectives. Improvement plans (tactics) prioritize processes (methods) for attention through policy management. Resources are allocated between methods by setting targets. An improvement plan usually follows a 1-to-2-year time horizon. The members of the PDC utilize the following steps to construct a set of integrated improvement plans.

Step 1a

Develop the corporate improvement plans needed to promote the corporate strategic objectives. "Gap analysis" is used to study the root cause(s) of the difference between customer and employee requirements, and organizational performance for each strategic objective.* The members of the PDC assign a group of staff personnel to study the gap for a particular strategic objective. For example, the group might study the gap over time

---

* These "gap analyses" are performed on the strategic objectives. The prior "gap analyses" were performed on the rows in the Table of Tables; see Appendix 6A.

and determine that it is stable and contains only common variation. Next, they construct a Pareto diagram of the common causes of the gap, isolate the most significant common cause, and develop a cause-and-effect diagram. Then the staff personnel study the relationship between the suspected root (common) cause and the size of the gap. If the staff personnel determines the relationship to be significant, they recommend a tactic for consideration as part of the organization's improvement plan to the members of the PDC.

The relationship between corporate strategic objectives and corporate improvement plans (tactics) can be seen in Figure 6.4. Relationships are measured on the following scale: 3 = strong relationship, 2 = moderate relationship, 1 = weak relationship, and blank = no relationship.

Every strategic objective should be adequately serviced by one or more tactics. If a strategic objective is not being serviced by any tactic (or not adequately serviced), one or more tactics are developed to service the strategic objective. All columns of the matrix in Figure 6.4 should contain at least one score of 2 or 3.

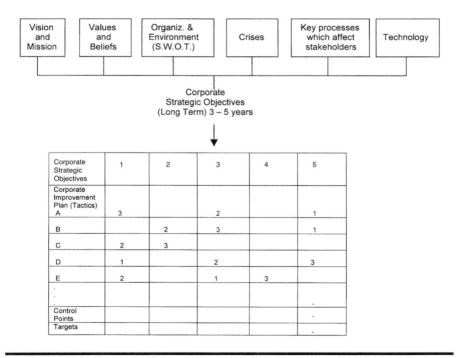

Figure 6.4 Relationship between Corporate Strategic Objectives and Corporate Improvement Plans

Step 1b

Develop the departmental improvement plans (tactics) needed to promote the departmental strategic objectives. Organizations need a mechanism for setting policy and allocating responsibility and resources in departments and divisions to promote corporate policy. Such a mechanism is shown in Figure 6.5.

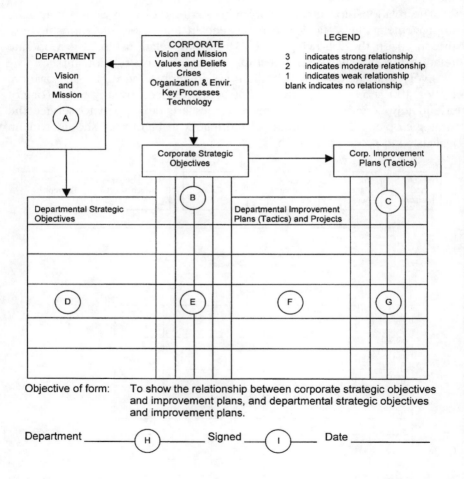

Figure 6.5 Setting Policy, and Allocating Responsibility and Resources

Source: Abstracted from Florida Power & Light Co., Description of Quality Improvement Program, Southern Division, 1988, p. 16.

Section A of Figure 6.5 provides departmental management an opportunity to create departmental vision and mission statements that promote the corporate vision and mission statements. Section B of Figure 6.5 lists the corporate strategic objectives (see the columns of Figure 6.4). Section C of Figure 6.5 lists the corporate improvement plans (tactics) (see the rows of Figure 6.4). Section D of Figure 6.5 list the departmental strategic objectives. Departmental management considers all the information utilized in developing the corporate strategic objectives when developing the departmental strategic objectives. Section E of Figure 6.5 shows the strength of relationships between corporate strategic objectives and departmental strategic objectives. Relationships are measured on the following scale: 3 = strong relationship, 2 = moderate relationship, 1 = weak relationship, and blank = no relationship. Every corporate strategic objective is adequately serviced by one or more departmental strategic objectives. If a corporate strategic objective is not being serviced, or adequately serviced, by any departmental strategic objectives, then one or more departments develop strategic objectives to service that corporate strategic objective. Section F of Figure 6.5 lists the departmental improvement plans (tactics) required to promote the corporate improvement plans (tactics). The departmental improvement plans result in quality improvement projects. When entering this information, it is important to line up departmental improvement plans with their corresponding departmental strategic objectives. Finally, section G of Figure 6.5 shows the strength of relationships between corporate improvement plans and departmental improvement plans. Relationships are measured on the following scale: 3 = strong relationship, 2 = moderate relationship, 1 = weak relationship, and blank = no relationship. Every corporate improvement plan is adequately serviced by one or more departmental improvement plans. If a corporate improvement plan is not being serviced or adequately serviced, by any departmental improvement plan, one or more departments develop improvement plans to service that corporate improvement plan. Section H of Figure 6.5 indicates the name of the department or division filling out the form. Section I of Figure 6.5 shows the signature of the departmental manager responsible for setting policy.

Step 2

The members of the PDC select and operationally define control points and targets for the corporate improvement plans (see Figure 6.4). Control points are measures about policies (results) that are managed with data. Targets are the desired (nominal) level for the control points set by management.

These control points and targets are deployed into departmental control points and targets. Targets are used to allocate resources for improvement projects to departments.

The members of the PDC ask the following questions:*

1. Have the control points used to monitor improvement plans been operationally defined?
2. Have targets been assigned to projects to optimize the corporate strategic objectives?

Step 3

The members of the PDC review and prioritize the projects called for in the improvement plans; see section F of Figure 6.5. The members of the PDC reach consensus on the priorities. They ask the following questions:**

1. Are priority projects well defined?
2. Will taking care of these priority projects help achieve the strategic objectives?
3. Are there better ways to achieve the strategic objectives?
4. Have the costs associated with pursuing the strategic objectives been studied?
5. Have the most appropriate projects been highlighted for study in the improvement plans?
6. Are projects defined with enough specificity so that everyone understands them?
7. Were projects discussed with relevant people and groups?
8. Were constraints on methods considered by the PDC?
9. Has the effectiveness of the projects been studied?
10. Are sufficient resources available for the projects?

Step 4

The members of the PDC communicate the projects that emanate from corporate and departmental improvement plans to the Local Steering Teams (LSTs), including the allocation of resources. The members of the PDC and an LST come to consensus on projects, targets, and resources in meetings called *catchball sessions.*

---

* These questions have been adapted from Mizuno, *Management for Quality Improvement,* p. 106.
** The questions in this section are paraphrased from Mizuno, *Management for Quality Improvement,* p. 106.

Local Steering Teams

The members of the LSTs are responsible for coordinating and carrying out the projects set out in the corporate and departmental improvement plans. The members of the PDC and LSTs should reach consensus on the priorities assigned to methods via targets that allocate resources. The members of LSTs ask the following questions:*

1. Are priority projects well defined?
2. Will taking care of these priority projects help achieve the improvement plans and strategic objectives?
3. Are there better ways to achieve the strategic objectives?

POLICY DEPLOYMENT

The strategic objectives and improvement plans are deployed by the members of the PDC through assignment of responsibility for action to people or groups of people in departments (see step 31 of the Detailed Fork Model in Figure 1.2). The assignment of responsibility is discussed by the members of the PDC and LST in meetings called *Mini-SITCONS*. Mini-SITCONS consider the costs to improve and innovate methods. After costs have been identified, it may be necessary to renegotiate project budgets. This process continues until all parties reach a consensual agreement on projects; again, this is called *catchball*. Finally, the members of the PDC ensure that the agreed-upon projects incorporate the information determined in the Table of Tables.

The assignment of responsibility for a project to a manager creates an opportunity to conduct a project that will result in improved or innovated "best practice" methods, the allocation of necessary resources, and an obligation to predict the contribution of projects to strategic objectives.

A tool that can be used to track the contribution of departmental improvement plans in the pursuit of a corporate objective is the *flag diagram*. A flag diagram** is used to monitor the contribution of departmental targets toward a global corporate target (see Figure 6.6). There are two types of flag diagram systems:

---

* The questions in this section are paraphrased from Mizuno, *Management for Quality Improvement,* p. 106.
** The material on the flag system was developed from the following resources: (1) Kano, N., *Second Report on TQC at Florida Power & Light Company* (Miami, FL), October 1, 1986, pp. 3 and 34. (2) Aldecocea, L., *A Flag System Application for Monitoring Timeliness of Installation for a Daily Process*, University of Miami (Coral Gables, FL), May 4, 1990.

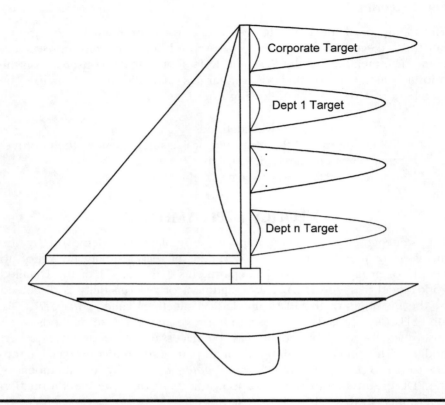

Figure 6.6 Generic Flag Diagram

■ *Additive flag system.* In some applications of the flag system, corporate targets are the summation of departmental targets. When targets are additive, it is easy to determine how to allocate resources between projects by targets. If all departments meet or exceed the targets pertaining to a particular corporate objective, the corporate objective(s) will be achieved.

■ *Non-additive flag system.* In other applications of the flag system, corporate targets are not the summation of departmental targets. When targets are not additive, it is difficult to determine how to allocate resources between projects by targets. If all divisions meet or exceed the targets pertaining to a particular corporate objective, it is unknown if the corporate objective will be achieved. In this case, knowledge and experience are required to determine the relationships between corporate objectives and their targets.

For example, progress toward a corporate objective of "continue to improve reliability of service to customers" could be tracked through the control point "number of minutes of downtime per year" into departments providing service to customers (Figure 6.7). An integrated set of project targets would be set (Figure 6.7) to allocate resources to the departments for projects to optimize the organization in respect to "continue to improve reliability of service to customers."

Deployment of an improvement plan project is completed when a project team has been assigned responsibility to improve or innovate a method. Figure 6.8 shows the projects, channels of communication, type of coordination, and resources necessary to implement policy in a department. Column A of Figure 6.8 shows departmental strategic objectives (see section D of Figure 6.5). Column B of Figure 6.8 shows the departmental improvement plans (tactics) needed to promote the departmental strategic objectives (see section F of Figure 6.5). Column C of Figure 6.8 shows the project(s) necessary for each departmental strategic objective. Each manager assigned a project must sign the appropriate line of column C indicating his or her acceptance of the project. Column D of Figure 6.8 shows the channels of communication and types of coordination between departments needed to carry out the projects shown in column C. The manager of each department named as a necessary supporter of a project must sign the appropriate line of column D, indicating his or her willingness to assist in the conduct of the project. Column E of Figure 6.8 shows the *additional* financial and human resources needed to carry out the projects shown in column C.

The members of an LST ask the following questions of all project team members:*

1. Are information channels between all relevant people and groups open to promote the improvement plan?
2. Has a detailed schedule been set up for carrying out the improvement plan?

POLICY IMPLEMENTATION

Policy is implemented in two ways (see step 32 of the Detailed Fork Model in Figure 1.2). First, policy is implemented when teams work on projects to improve and/or innovate processes. Second, policy is implemented when

---

* The questions in this section are paraphrased from Mizuno, *Management for Quality Improvement,* p. 106.

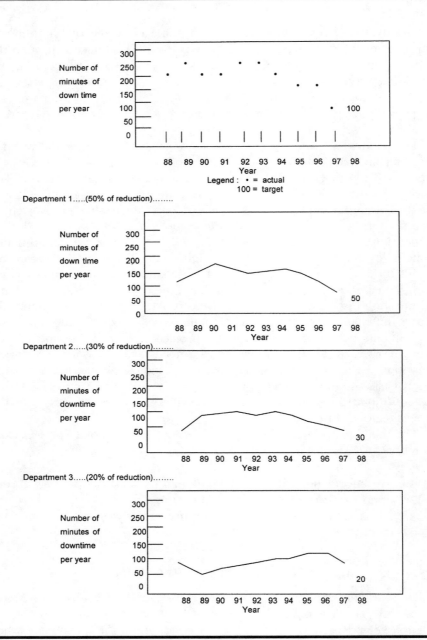

Department 1.....(50% of reduction)........

Department 2.....(30% of reduction)........

Department 3.....(20% of reduction)........

Figure 6.7 Fictional Example of a Flag Diagram.
Corporate Improvement Plant (Tactic): "Continue to improve reliability of service to customers."
Corporate Control Point: "Number of minutes of down time per year"

Departmental Strategic Objectives	Departmental Improvement Plan (Tactics)	Project	Other Depts. Affected/ Coordination Needed	Resources • Dollars • People
(A)	(B)	(C)	(D)	(E)

Figure 6.8 Implementation of Policy in a Department

Source: Florida Power & Light Co., 1990.

departments use the revised processes and measure their results in respect to improvement plans and strategic objectives.

The members of the EC assign responsibility for the promotion of each strategic objective to a high-level executive. The executive removes any impediments to progress for his or her strategic objective. Furthermore, the executive coordinates efforts in respect to the strategic objective throughout the organization. Finally, the members of the PDC, in conjunction with the members of various LSTs, may have to modify improvement plans as they proceed over time.

POLICY FEEDBACK AND REVIEW

Periodic management reviews are conducted at two levels (see step 33 of the Detailed Fork Model in Figure 1.2). First, the members of the EC review progress toward each strategic objective (and its improvement plans) monthly. Brief presentations for each strategic objective are made by the high-level executive responsible for that strategic objective. Twice a year, each strategic objective is selected for a detailed management review. The detailed review probes very deeply into the issues surrounding a strategic objective. The members of the EC insist that all process modifications are supported by sound analysis. This may require that action items be assigned to the high-level executive responsible for a strategic objective. The members of the EC also follow up on the action items. Second, the members of the PDC and of appropriate LSTs review progress for each project. The purpose of these reviews is to provide feedback to project team members that promotes process improvement efforts. The members of the PDC make sure that all process modifications are supported by sound analysis of data. This may require that action items be assigned to the members of a project team. The members of the PDC also follow up on the action items.

The members of the EC, PDC, or appropriate LSTs ask the following questions of project team members:

1. Does your organization have a vision statement and a mission statement?
2. Do you know what they are?
3. Do you understand how you can contribute to the vision and mission of your organization? How do you know?
4. Does your department have vision and mission statements?
5. Do you know what they are?
6. Do you understand how you can contribute to the vision and mission of your department? How do you know?
7. Do you understand the methods by which you will achieve the vision and mission of your department? Organization?
8. Do you know which methods are most critical to pursue the vision and mission of your department? Organization?
9. Do you know the aims of these methods?
10. Do the aims of these methods support the aims of your department? Organization?
11. Are these methods necessary?
12. What are the critical control points for these methods?
13. Have the critical control points been operationally defined?
14. Have you used the SDSA cycle to standardize methods?
15. Have you used the PDSA cycle to improve and innovate methods?
16. Do you understand who the customers of these methods are?
17. Do you understand the needs of those customers?
18. Do you know who the suppliers of these methods are?
19. Do you understand the needs of those suppliers?
20. Do you understand how these methods interact with other methods in your department? Organization? How do you know?
21. Have you been trained in team skills and basic quality improvement tools?
22. Have you received training in the methods critical to your job?

23. Is your training in job skills updated as your job changes over time?
24. Have training manuals been updated as jobs change over time?
25. Do you receive feedback on the performance of the methods with which you work on a continuous basis?
26. Do you feel ownership of the methods with which you work?
27. Do you take pride in the outcome of your work?
28. Do you take joy in your work?
29. Are you an empowered employee? How do you know?
30. Does your supervisor lead you in the conduct of planned experiments aimed at improvement and innovation of methods?
31. Do you have latitude to modify the methods you use on your job to take advantage of your unique skills and abilities?
32. Do you need the latitude you have in respect to a method?
33. Can all of your colleagues who perform a particular method produce equal outcomes? How do you know?
34. Do you trust your supervisor to support the decisions you make within the latitude given to you in respect to a particular method?
35. Is your supervisor working toward eliminating fear in your department? Organization? How do you know?
36. Are you implementing the improvement plan and/or projects per schedule?
37. Are records being kept of quality improvement efforts?
38. Are you revising the improvement plan as necessary?

The members of the EC and PDC ask the following questions of themselves:

1. Are we effectively conducting management reviews?
2. Are we improving and innovating the management review process?
3. Are methods being standardized and revised as required by the improvement plan?

PRESIDENTIAL REVIEW

Finally, the President conducts the Presidential Review (see step 34 of the Detailed Fork Model in Figure 1.2) of the major areas within the organization. Department managers present their efforts using the Quality Improvement story format. The purpose of the Presidential Review is to collect information used to establish the quality strategy and goals of the organization and to determine progress toward presidential policy. Presidential Reviews provide input for setting policy for the following year (see step 29 of the Detailed Fork Model in Figure 1.2). See the section of this chapter on "Initial Presidential Review" for details. The management system is improved with each successive policy management cycle.

FLOWCHART OF POLICY MANAGEMENT

An integrated flowchart depicting the relationship between the PDSA cycle and the five steps of policy management, and corporate and department responsibility for policy management, is shown in Figure 6.9.

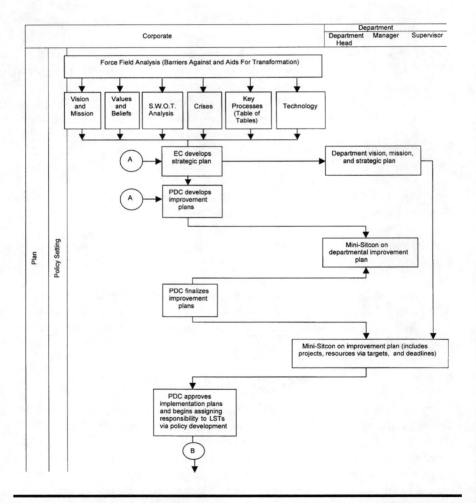

Figure 6.9 Integrated Flowchart of the Relationships between the PDSA and Policy Management

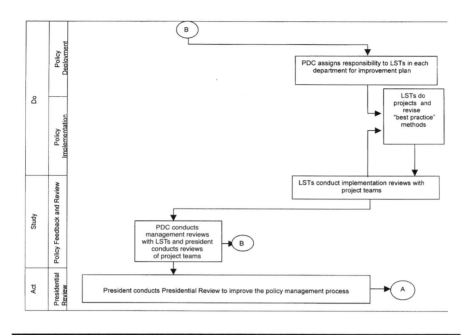

Figure 6.9 (continued) Integrated Flowchart of the Relationships between the PDSA and Policy Management

RELATIONSHIP BETWEEN POLICY MANAGEMENT AND DAILY MANAGEMENT

The relationship between policy management and daily management can be understood by viewing an organization as a tree. The vision is the root system, the mission is the trunk, the strategic objectives yield the major branches, and the improvement plans are smaller branches emanating out of the strategic objective branches. As expertise is developed with policy management methods, they are moved into daily management methods.

Frequently, employees claim they do not have time for policy management due to the demands of their daily routine. Doing daily management and cross-functional management removes non-value-added daily routine to free up time for policy management.

PERSONAL EXAMPLE OF POLICY MANAGEMENT

This section presents the application of policy management to a person's life. All the steps of policy management are used in this practical example, which demonstrates how to implement this detailed procedure. Bart is a 40-year-old manager in a large company. For the most part, his life has gone according to his plans. He is a respected manager, earns a comfortable salary, and is reasonably happy in his personal and family life. As he enters his forties, Bart begins to wonder whether he can improve his situation. Stimulated by what he is involved in at work, he decides to apply the principles of quality management to his life. Bart's application of policy management follows.

Policy Setting

Policy setting is illustrated below.

Pros and Cons to Transformation

Bart performs a force field analysis and asks himself the following question: "Do I have the energy necessary to do Quality Management in my life?" From the force field analysis, he finds the "forces for change" to be more compelling than the "forces against change" (see Figure 6.10). Bart decides to apply policy management in his life.

Forces for Change	Forces Against Change
Some dissatisfaction with my current life situation	No time to establish a policy management plan
Complaints from spouse, daughter, and brother	Family, friends, and colleagues will think I am a lunatic
Need to be more efficient and productive at work	

Figure 6.10 Force Field Analysis

Values and Beliefs

Bart feels that the values and beliefs embodied in Dr. Deming's theory of management are completely consistent with his own, so he adopts these values and beliefs. They are (1) manage to optimize the entire system, (2) manage to create a balance of intrinsic and extrinsic motivation, (3) manage with a process and results orientation, and (4) manage to promote cooperation.

Vision and Mission

Bart's personal vision and mission define his future desired state and his current reason for existence.

> *Personal Vision:* To be at peace with my world and to generate positive energy into the universe.
> *Personal Mission:* To continuously improve my mind, body, and relationships.

Organizational and Environmental Factors

Bart conducts a S.W.O.T. analysis of himself. An abbreviated listing of strengths, weaknesses, opportunities, and threats is given below.

> *Strengths:* honest, forthright, mature, intelligent, excellent educational background, inquisitive, considerate, willing to help others with their problems, physically healthy, excellent communicator, strong support system of family and friends, well-established and ingrained values and beliefs, financially secure, professionally secure, able to stay focused, capable of retaining many facts until a strategy emerges, open to new ideas, resources available for self-improvement, …
> *Weaknesses:* high cholesterol, 25 pounds overweight, unilingual, resents working in nonproductive groups, not good at understanding other people's viewpoints, unable to change own behavior, poor short-term memory, low frustration tolerance, …
> *Opportunities:* excellent business contacts, professional opportunities expanding, excellent system of personal mentors, increased ability to travel, …
> *Threats:* increased demands to spend time working at the expense of personal life, parents passing away and not being prepared for the loss, the unknown and unknowable risks of life,…

Synthesis

Bart studies his S.W.O.T.s and looks for Strengths and Opportunities that overcome his Weaknesses and Threats. He decides that his "available time for" and "ability to understand the benefits of" a healthy lifestyle will help him overcome his physical and emotional condition.

Crisis

The crisis known to Bart is diminution in quality of life due to moderate swings in mood (too much common variation in a stable system).

Key Processes

"Voice of the Customer" and "Voice of the Business" analyses help prioritize processes (methods) for improvement attention. The purpose of these analyses is to provide customer and personal input into the determination of strategic objectives.

Voice of the Customer

Bart surveys each of his stakeholders and asks them to answer the following question: "From your perspective, what should I focus on to pursue my mission statement?" Data are collected from each individual. All issues that emerge from this analysis are quantified on a "dynamite" scale; the scale ranges from 1 stick of dynamite (unimportant or done well) to 5 sticks of dynamite (very important and not done well). The dynamite scale was developed to create data that can be averaged.* The "dynamite" data is being used to impart a "feel" for which issues are priority issues in Bart's pursuit of his mission. The "dynamite" data was analyzed using Pareto-type analysis.

> *Spouse:* The spouse's survey indicates 17 separate items of concern, which are grouped into four subcategories: "Child care responsibility," "Home care responsibility," "Finances," and "Time allocation." All 17 items are rated on the dynamite scale (Figure 6.11). An analysis of the priority rankings shown in Figure 6.11 indicates that 36.6% of the spouse's issues involve household chores: "shop at supermarket," "shop at drug store," and "go to the dry cleaners" (Figure 6.12). Therefore, "Home care responsibility" is the spouse's critical area of concern.
>
> *Child:* The child's survey indicates three items of concern (Figure 6.13). An analysis of the child's issues indicates that "increase play time" is the most critical area of concern; it accounted for 44% of the child's issues (Figure 6.14).
>
> *Parents and Brother:* The parents' and brother's survey indicates several items of concern (Figure 6.15). An analysis of the parents'

---

* The author realizes that he is taking liberties with the scale of the data.

Category	Customer Ranking Survey Results (dynamite scale)
(a) Child care responsibility	
1. Drive child to/from school	1
2. Prepare child's dinner	2
3. Care for ill child	2
4. Help child with homework	1
(b) Home care responsibility	
1. Shop at supermarket	5
2. Shop at drug store	5
3. Go to dry cleaner	5
4. Sort clothes for wash	1
5. Make bank deposits	4
(c) Finances	
1. Prepare household accounts	1
2. Generate family income	1
3. Manage expense control	1
(d) Time allocation	
1. Decrease work time	4
2. Increase family time	2
3. Increase play time with child	2
4. Decrease business travel	2
5. Increase pleasure travel	2

Figure 6.11 Voice of Customer for Spouse

and brother's concerns shows that 45% of all issues directly involve the brother (Figure 6.16). The brother's three issues are "share private feelings and dreams," "increase private time together," and "support each other."

Voice of the Business

The Voice of the Business seeks to answer the following question: "What should I focus on to pursue my mission?" Bart compiles a list of 11 issues that are critical to the fulfillment of his mission (see Figure 6.17). These 11 issues are rated in respect to severity, urgency, trend, and importance to customer (see the prioritization matrix in Figure 6.18). The ratings are

Category	Rank	%	Cum. %	
Shop at supermarket	5	12.2	12.2*	36.6% of
Shop at drug store	5	12.2	24.4*	spouse's issues
Go to dry cleaner	5	12.2	36.6*	involve household
Decrease work time	4	9.6	46.2	chores
Make bank deposits	4	9.6	55.8	
Increase family time	2	4.9	60.7	
Increase play time with child	2	4.9	65.6	
Decrease business travel	2	4.9	70.5	
Increase pleasure travel	2	4.9	75.4	
Prepare child's dinner	2	4.9	80.3	
Care for ill child	2	4.9	85.2	
Drive child to/from school	1	2.45	87.65	
Help child with homework	1	2.45	90.10	
Sort clothes for wash	1	2.45	92.55	
Prepare household accounts	1	2.45	95.00	
Generate family income	1	2.45	97.45	
Manage expense control	1	2.45	99.90	
TOTAL..............................41		99.90		

Figure 6.12 Analysis of Spouse's Voice of the Customer Data

Category	Customer Ranking Survey Results (dynamite scale)
Child	
(a) Increase play time	4
(b) Relax sleep over rules	2
(c) Decrease food restrictions	3

Figure 6.13 Voice of the Customer for Child

then multiplied to obtain a total score for each issue. This method is used just to demonstrate an alternative to the dynamite scale used in the "Voice of the Customer" analysis.

The items in the matrix are arranged in descending order of priority (Figure 6.19). An analysis of the data in Figure 6.19 is shown in Figure 6.20. Four of the eleven issues account for 90.7% of Bart's issues. These

Category	Rank	%	Cum. %	
Increase play time	4	44	44*	44% of
Decrease food restrictions	3	33	77	child's issues
Relax sleep over rules	2	23	100	involve play
TOTAL................................	9	100		

Figure 6.14 Analysis of Child's Voice of the Customer Data

Category	Customer Ranking Survey Results (dynamite scale)
(a) Father	
1. Increase private time together	3
2. Go to sporting events together	2
3. Work on car together	2
4. Work on garden together	2
(b) Mother	
1. Determine common interests	4
2. Increase private time together	3
(c) Brother	
1. Increase private time together	4
2. Share private feelings and dreams	5
3. Learn to support each	4

Figure 6.15 Voice of the Customer for Parents and Brother

Category	Rank	%	Cum.	%
Brother – Share private feelings and dreams	5	17	17*	45% of
Brother – Increase private time together	4	14	31*	issues
Brother – Learn to support each	4	14	45*	involve
Mother – Determine common interests	4	14	59	commu-
Mother – Increase private time together	3	10	69	nication
Father – Increase private time together	3	10	79	with
Father – Go to sporting events together	2	7	86	brother
Father – Work on car together	2	7	93	
Father – Work on garden together	2	7	100	
TOTAL...................................	29	100		

Figure 6.16 Analysis of Parents' and Brother's Voice of the Customer Data

1) Increase pleasure travel with spouse
2) Increase private study time
3) Maintain study group time
4) Increase play time with child
5) Increase fun time with friends
6) Increase exercise time
7) Decrease body weight
8) Plan and shoot fireworks displays
9) Plan and execute special events
10) Increase private time with spouse
11) Improve financial security

Figure 6.17 Voice of the Business

Category	Severity	Urgency	Trend	Import to Customer	TOTAL SCORE
(1) Inc. travel with spouse	1	1	2	3	6
(2) Inc. private study time	1	1	4	5	20
(3) Mntn study group time	1	1	2	5	10
(4) Inc. play with child	3	3	4	4	144
(5) Inc. fun with friends	1	1	2	3	6
(6) Inc. exercise time	3	3	3	5	130
(7) Dec. body weight	4	4	4	4	256
(8) Fireworks displays	1	1	2	1	2
(9) Special events	1	1	3	3	9
(10) Inc. time with spouse	2	2	4	5	80
(11) Imp. financial security	1	1	3	3	9

Figure 6.18 Prioritization of Matrix of the Voice of the Business

include "decrease body weight," "increase play time with child," "increase exercise time," and "increase time with spouse."

Table of Tables

The prioritized issues from the "Voice of the Customer" and "Voice of the Business" studies are combined into one prioritized list in the rows of Figure 6.21. The processes that form Bart's life are listed in the columns of Figure 6.21.

Category	Severity	Urgency	Trend	Importance to Customer	TOTAL SCORE
(7) Dec. body weight	4	4	4	4	256
(4) Inc. play with child	3	3	4	4	144
(6) Inc. exercise time	3	3	3	5	130
(10) Inc. time with spouse	2	2	4	5	80
(2) Inc.private study time	1	1	4	5	29
(3) Mntn. study group time	1	1	2	5	10
(9) Special events	1	1	3	3	9
(11) Imp. financial security	1	1	3	3	9
(1) Inc. travel with spouse	1	1	2	3	6
(5) Inc. fun with friends	1	1	2	3	6
(8) Fireworks displays	1	1	2	1	2

Figure 6.19 Ranking of Voice of the Business Data

Category	Total Score	%	Cum. %
(7) Dec. body weight	256	38.1	38.1*
(4) Inc. play with child	144	21.4	59.5*
(6) Inc. exercise time	130	19.3	78.8*
(10) Inc. time with spouse	80	11.9	90.7*
(2) Inc. private study time	20	3.0	93.7
(3) Mntn. study group time	10	1.5	95.2
(9) Special events	9	1.3	96.5
(11) Imp. financial security	9	1.3	97.8
(1) Inc. travel with spouse	6	0.9	98.7
(5) Inc. fun with friends	6	0.9	99.6
(8) Fireworks displays	2	0.3	99.9
TOTAL	672	99.9	

Figure 6.20 Analysis of the Voice of the Business Data

The *importance to customer or employee* scale (see column "A" in Figure 6.21) is ranked from 1 (unimportant to customer or employee) to 5 (important to customer or employee). The rankings are developed by company personnel and are modifications of the rankings identified in the Voice of the Customer and Voice of the Business studies. The scale indicates that "decrease body weight" (5), "increase exercise time" (4),

Legend:

	Score
◎ Strong Relationship	3
O Moderate Relationship	2
△ Weak Relationship	1
Я No Relationship	0

Needs & Wants	Fitness			Occupation		Leisure		Learning		Spiritual	Psychological			Relationships				Importance to customer or employee (A)	Current level or importance to customer or employee (B)	Desired level of performance by management (C)	Total Weight = A(C/B)
	Weight	Diet (Food Intake)	Exercise	Joy in Work	Revenue	Travel	Entertainment	Education	Training		Self Esteem	Well being	Joy	Love	Communication	Fun	Sharing responsibility				
Decreased body weight - VOB	◎	◎	◎					△			△	△						5	1	5	25
Increase exercise time - VOB	◎		◎									△						4	1	5	20
Spouse-Household chores - VOC (Supermarket, Drug Store, Dry Cleaner)													△				◎	5	1	5	25
Daughter - Play time - VOB & VOC (Increase play time with child)							△				△		O	◎	◎	◎	O	4	3	2	2.67
Brother - Increase communication - VOC (Feelings, Time, Support)													O	△	◎			1	3	2	0.67
	135	75	135	0	0	0	2.67	25	0	0	27.67	45	31.68	8.68	10.02	8.01	80.34				

Figure 6.21 Table of Tables

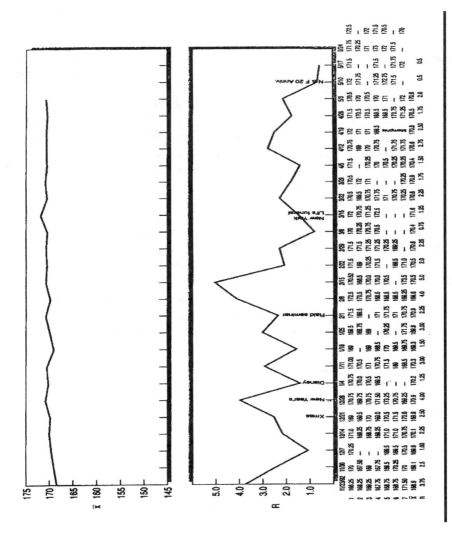

Figure 6.22 Gap Analysis Decrease Body Weight

"household chores" (5), and "increase playtime with child" (4) are important to customers and Bart.

The *current level of performance to customers or employees scale* (see
column "B" in Figure 6.21) is ranked from 1 (large gap between customer
or employee desired requirements and actual performance) to 5 (small
gap between customer or employee desired requirements and actual
performance). The rankings are developed by company personnel. The
gap analyses of "decrease body weight" (see Figure 6.22), "increase
exercise time (see Figure 6.23), and "household chores" (see Figure 6.24)
all indicate large gaps between desired and actual performance, in the
opinion of the company personnel. Specifically, the gap analysis of
"decrease body weight" (see Figure 6.22) shows that Bart's average body
weight has steadily increased to 172 pounds. According to Bart's physician,
he should weigh 145 pounds. Additionally, the gap analysis of "increase
exercise time" (see Figure 6.23) shows that Bart has not been capable of
maintaining the required three exercise periods per week. He achieved
three exercise periods per week in only 25% of the weeks in the trial
period. Finally, gap analysis of "household chores" (see Figure 6.24) shows
that Bart has not been capable of maintaining the necessary one trip per
week to the supermarket, drug store, and dry cleaner most of the time
in the trial period.

The *desired level of performance by management* scale (see column
"C" in Figure 6.21) is ranked from 1 (improvement is not urgent) to 5
(improvement is urgent). Again, the rankings are developed by company
personnel based on management's view of the urgency to improve voice
of the customer or voice of the business items. The scale indicates that

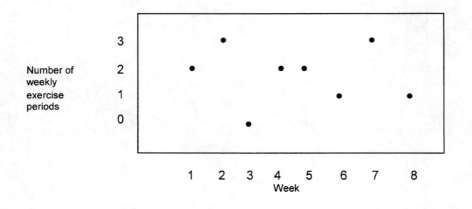

Figure 6.23 Gap Analysis Exercise Time

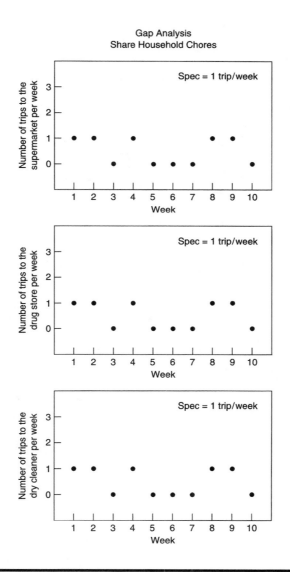

Figure 6.24 Gap Analysis Household Chores

"decrease body weight," "increase exercise time," and "household chores" are all deemed in need of urgent improvement attention by Bart.

The *total weight* scale (see column "A(C/B)" in Figure 6.21) is computed by multiplying the *importance to customer or employee* scale by the *desired level of performance by management* scale to obtain a significance indicator (1 signifies extreme unimportance and 25 indicates extreme importance). This significance indicator is divided by the *current level of performance scale*. The resulting number is the *total weight* scale. A *total weight* of 0.2 ([1 × 1]/5) indicates a very unimportant voice of the customer or voice of the business item, while a *total weight* of 25 ([5 × 5]/1) indicates an extremely important voice of the customer or voice of the business item. The *total weight* scale shows the need to focus attention on "decrease body weight" (total weight = 25), "increase exercise time" (total weight = 20), and "household chores" (total weight = 25) in Bart's strategic objectives, see Figure 6.21.

The cells of Figure 6.21 show the strength of the relationships between "Voice of the Customer" and "Voice of the Business" issues and the processes that form Bart's life. A doughnut symbol indicates a strong relationship (value = 3), a circle indicates a medium relationship (value = 2), a triangle indicates a weak relationship (value = 1), and a blank indicates no relationship (value = 0). The cell relationships are determined by company personnel familiar with the processes listed in the columns, and issues listed in the rows, of the Table of Tables. A numeric quantity is computed for each cell by multiplying the value for its symbol by the *total weight* of its row. For example, the value for the cell defined by the "fitness — weight" column and the "decrease body weight" row is 3. The *total weight* of its row is 25. Hence, the numeric quantity computed for the cell is 75 (3x25). This operation is done for each cell in the Table of Tables. Finally, the numeric quantities are summed up for each column.

The "Table of Tables" reveals that "weight" (priority = 135), "exercise" (priority = 135), and "shared responsibilities" (priority = 80.34) are the processes that must receive attention in Bart's strategic objective if he is to surpass the needs and wants of his customers and himself (see the bottom row of Figure 6.21).

Technology

Bart realizes that using a treadmill at home while watching television, as opposed to jogging outside, makes him want to exercise more.

Develop Strategic Objectives

Bart's strategic objectives are

1. To become healthy.
2. To continually improve personal relationships.

They emerge from an analysis of Bart's vision and mission statements, his values and beliefs, a S.W.O.T. analysis of his life, the crises Bart faces, the key processes that have to be improved to delight Bart's stakeholders, and technological issues that might affect Bart's life. The best Quality Management method for generating strategic objectives is a thoughtful, yet subjective, analysis of all inputs by the members of the Executive Committee. It is critical that all stakeholders agree to support the strategic objectives.

Develop the Improvement Plans

Bart brainstorms a list of the barriers that prevent him from becoming healthier (*strategic objective 1*) as part of a "gap" analysis. The list includes the following potential causes: too much body weight, too little exercise, unbalanced diet, too much stress, and too little sleep. An analysis of the interrelationships between the potential causes indicates that "too little exercise" is the root cause which affect poor health in Bart's life (see Figure 6.25).

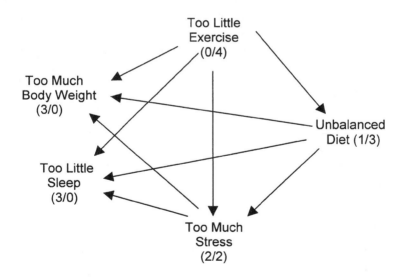

Figure 6.25 Interrelationship Diagraph Causes of Diminished Health

The arrows in between the items on the interrelationship diagraph indicate the direction of "cause to effect" relationships, for example, "too much stress" partially causes "too little sleep," while "too much stress" is partially caused by (the partial effect of) "too little exercise." The numbers in parentheses under each item on the interrelationship dagraph, such as (1/3) under "unbalanced diet," indicate the number of arrows entering (effecting) an item and leaving (causing) another item, respectively. The item with the largest number of arrows entering it is called the "root effect" item because it is effected by so many other items. Root effect items are frequently viewed as the sources of problems, but since they are effected by so many other items, it is difficult to do anything about them. "Too much body weight" and "too little sleep" are the root effect items in this interrelationship diagraph. They make sense in that these are commonly the things people worry about in respect to health, but they are difficult to do anything about because they are effected by so many other things. On the other hand, "too little exercise" is the "root cause" item in this interrelationship diagraph; it effects (causes) the most other items. This is the improvement tactic Bart should focus on to pursue his first strategic objective.

Bart collects frequency data from his diary on the barriers that prevent him from improving his personal relationships for a 2-month period, as part of a "gap" analysis. The barriers include the following potential causes: failure to share responsibilities (house care, child care, time allocation), moderate swings in mood, not good at understanding other people's viewpoints, poor short-term memory, low frustration tolerance, and increased demand to spend time at work. Figure 6.26 shows the frequency of occurrence of each barrier, extracted from Bart's diary, and that failure to share house care responsibilities is the most significant cause that affects Bart's personal relationships.

Cause of poor relationships	#	%	Cum. %
responsibilities	52	72	72
mood	4	6	78
viewpoints	1	1	79
memory	2	3	82
frustration	12	17	99
time	1	1	100
Total	72	100	

Figure 6.26 Pareto Analysis of Personal Relationships

The above "gap analyses" indicate that Bart needs to construct improvement plans to deal specifically with developing an exercise regimen (tactic for strategic objective 1) and sharing house care responsibilities (tactic for strategic objective 2), if he wants to surpass his strategic objectives.

Bart operationally defines control points and establishes targets for an exercise regimen and house care responsibilities; in particular, going to the supermarket, drug store, and dry cleaner. The purpose of the targets is to allocate the resources available to Bart between his improvement plan tactics. In this case, the targets are based on medical science and years of experience with the chores. Control points and targets are shown in Figure 6.27.

Strategic Objective	Improvement Plan (Tactics)	Control Point	Target
To become healthy	develop an exercise regimen	# of exercise periods/week	3
To continually improve personal relationships	do supermarket shopping	# trips/week	1
	do drug store shopping	# trips/week	1
	go to dry cleaner	# trips/week	1

Figure 6.27 Control Points and Targets

Policy Deployment

Responsibility for the two projects discussed in policy setting are assigned to Bart. The following section presents the methods that will be deployed in his improvement plan.

An Exercise Regimen

Bart's method for improving his exercise regimen was discussed in Daily Management, Chapter 4.

House Care Responsibilities

Plan (developing a plan) — An important process in Bart's life highlighted for attention is "sharing responsibility" with his spouse, especially household chores such as shopping at the supermarket, shopping at the drug store, and picking up and dropping off clothes at the dry cleaner. Bart collects data on each of these three activities (methods) for 10 weeks. The run charts are shown in Figures 6.28, 6.29, and 6.30.

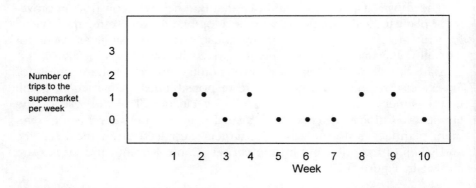

Figure 6.28 Run Chart Number of Trips per Week to the Supermarket

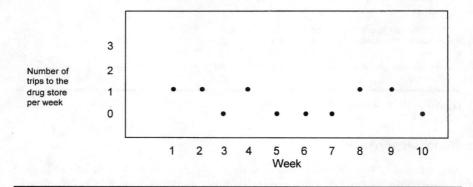

Figure 6.29 Run Chart Number of Trips per Week to the Drug Store

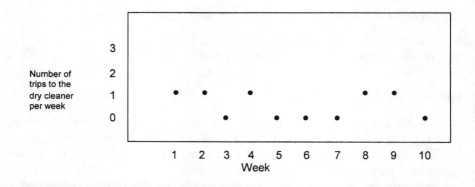

Figure 6.30 Run Chart Number of Trips per Week to the Dry Cleaner

Analysis of the run charts in Figures 6.28, 6.29, and 6.30 leads Bart to the realization that he needs to develop methods for "sharing responsibilities." He develops the flowcharts (methods) shown in Figures 6.31, 6.32, and 6.33.

Do (implementing the plan) — Bart records the number of trips to each location for 10 weeks after development of the methods. The data are shown in Figures 6.34, 6.35 and 6.36.

Study (checking the effectiveness of the plan) — The records show that all three targets are being met using the above methods.

Act (act) — Action to modify current method is taken if a trend develops indicating failure of the methods to meet targets. Continuous monitoring and documentation of the causes for additional trips will help modify the current practices and assure achievement of targets.

Policy Implementation

All methods in the improvement plan are implemented and monitored on a weekly basis: exercise regimen, trips to the supermarket, trips to drug store, and trips to dry cleaner. (*Note:* the exercise regimen is discussed in Daily Management, Chapter 4.) All methods are yielding predicted results for a 10-week period (see Figures 6.37, 6.38, 6.39, and 6.40).

Policy Study and Feedback

All methods implemented in the improvement plan are subject to monthly management reviews by Bart. He finds that methods are yielding predicted results. A year-end management review will be conducted at the appropriate time to determine if methods remain effective in respect to optimization of Bart's interdependent system of stakeholders.

Presidential Review

Presidential Review in the case study of personal policy management is equivalent to the yearly management reviews conducted by Bart of his strategic and improvement plans.

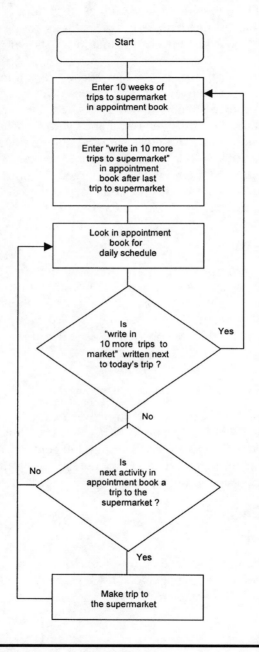

Figure 6.31 Method for Shopping at Supermarket

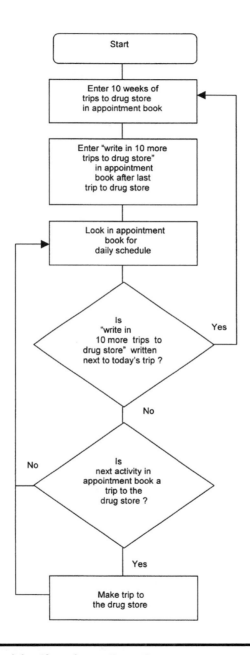

Figure 6.32 Method for Shopping at Drug Store

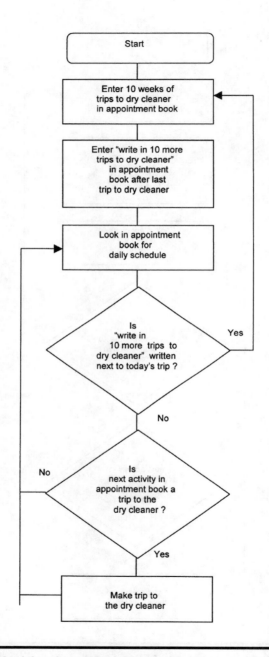

Figure 6.33 Method for Shopping at Dry Cleaner

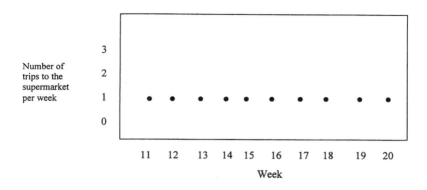

Figure 6.34 Continuation of Run Chart Number of Trips per Week to the Supermarket

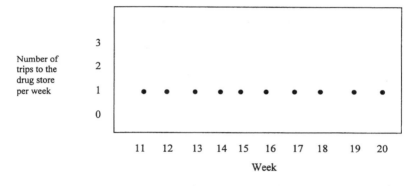

Figure 6.35 Continuation of Run Chart Number of Trips per Week to the Drug Store

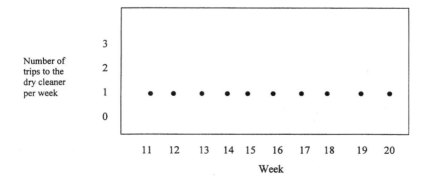

Figure 6.36 Continuation of Run Chart Number of Trips per Week to the Dry Cleaner

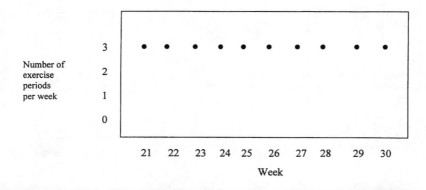

Figure 6.37 Continuation of Run Chart Number of Exercise Periods per Week

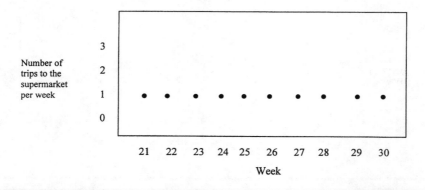

Figure 6.38 Continuation of Run Chart Number of Trips per Week to the Supermarket

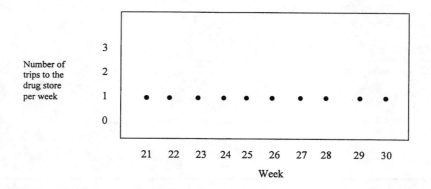

Figure 6.39 Continuation of Run Chart Number of Trips per Week to the Drug Store

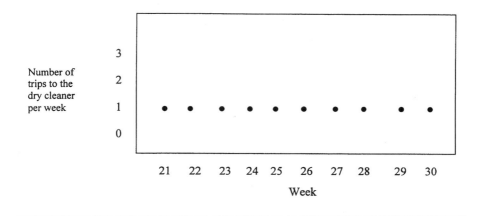

Figure 6.40 Continuation of Run Chart Number of Trips per Week to the Dry Cleaner

BUSINESS EXAMPLE OF POLICY MANAGEMENT*

Introduction

Florida Power & Light Co. (FP&L) is the largest utility furnishing the generation, transmission, distribution, and sale of electricity in the state of Florida. From its inception in 1925, FP&L has experienced steady growth. In November 1989, Florida Power & Light Co. achieved international recognition when its employees challenged for, and won, the prestigious Deming Prize for quality. The Deming Prize is awarded annually by the Japanese Union of Scientists and Engineers (JUSE) to companies that excel in the practice of the Japanese method of Quality Management. FP&L was the first non-Japanese company to win the Deming Prize.

Reasons for the Quality Improvement Program

By the early 1980s, FP&L was facing a hostile environment created largely by high inflation, decreasing customer sales, rising electric rates, and increasing fuel oil prices. The price of electricity was increasing faster than the Consumer Price Index (CPI). At the same time, competitive market

* *Source:* Gitlow, H. and Loredo, E., "Total Quality Management at Florida Power & Light Company: A Case Study," *Quality Engineering*, vol. 5, no. 1., 1992–1993, pp. 123-158. Additionally the author would like to thank J. Michael Adams, Corporate Manager of Quality Services, Florida Light & Power Company/FPLGroup, Inc., for his significant effort in the development of this business example.

pressures were beginning to affect FP&L's long-term prospects. Customer dissatisfaction grew as FP&L failed to meet increasing expectations for reliability, safety, and customer service. In the meantime, FP&L's inability to react quickly to new environmental demands added to their plight.

Objectives of the Quality Improvement Program

It was clear to FP&L's leadership that their existing managerial structures were not keeping pace with FP&L's rapidly changing internal and external environments. Above all, they recognized that FP&L was facing four significant crises that warranted complete restructuring of their managerial systems.

Crisis 1. Internal and external environments were changing faster than FP&L could adapt.

Crisis 2. Declining customer confidence and satisfaction.

Crisis 3. Uncertainty of the future of nuclear power supply.

Crisis 4. Price of electricity increasing faster than the Consumer Price Index.

As described by FP&L's top executives, crisis 1 was an internal issue that involved changing FP&L's corporate culture. FP&L had to change its mode of thinking from a supply-oriented mindset to a customer-oriented mindset, from a power generation company to a customer service company. To accomplish this, FP&L's managers needed a process that would allow them to identify and address the key issues surrounding customer satisfaction.

Crises 2, 3, and 4 were related to external issues. FP&L's leadership realized that these issues required systematic and resourceful management, guided by a vision, a mission, and strong strategic and business plans.

The first step toward resolving these crises was the promotion of a new management system. Prior to 1985, management by objectives (MBO) had been FP&L's principal policy for setting and achieving corporate goals. However, company top management had concluded that MBO was not capable of resolving the issues at the core of the four crises with which the company was to grapple. In particular, MBO's focus on the company's point of view rather than on the customer's needs conflicted with the establishment of a new corporate culture and subsequently, with the resolution of crisis 1. Also, MBO did not provide a systematic method for measuring and achieving corporate objectives. A system was needed that would be responsive to changes in FP&L's operating environment, as well as to the needs of their customers. To this end, FP&L established its Quality Improvement Program (QIP).

Reacting to the Crises

Determining Customer Needs

To a great extent, the success of FP&L's new management process (QIP) would depend on the development and application of a tool that could systematically identify and prioritize the needs of FP&L's customers. FP&L's diverse customer base is composed of direct customers (residential, commercial, and industrial users) and indirect customers (regulatory and governmental agencies, for example, the Nuclear Regulatory Commission, the Florida Public Service Commission, environmental and regulatory agencies, state and local governments, and the Federal Energy Regulatory Commission).

The tool FP&L developed to understand the "Voice of the Customer" for each of its market segments is the "Table of Tables" (Figure 6.41). It is used to build customer needs into FP&L processes. The Table of Tables shows the prioritized concerns of FP&L's diverse customer segments with regard to "sales and service quality," "delivery," "safety," "cost," and "corporate responsibility." By applying the Table of Tables, FP&L was able to develop strategies to resolve its four crises. For example:

1. Crises 1 and 3 were partially resolved by increasing the level and value of communications between FP&L and its operating environment.
2. Crises 2 and 4 were partially resolved by improving sales and service processes, which lowered costs and allowed FP&L management to decrease the price of electricity.

The development of the Table of Tables was the key to identifying and prioritizing customer needs and wants. In turn, this information was used to resolve internal and external crises by establishing corporate responsibility (e.g., for process improvement action), which was disseminated throughout the organization via policy management.

Establishing Divisional Objectives

Prior to FP&L's Quality Improvement Program (QIP), each division pursued its own vision and key objectives based on its unique set of characteristics (for example, see the differences between divisions in Figure 6.42b). More often than not, divisional objectives were optimized ahead of corporate objectives. This made it difficult to coordinate divisional objectives within the context of corporate policy and led to interdivisional rivalries and suboptimization of the corporate whole.

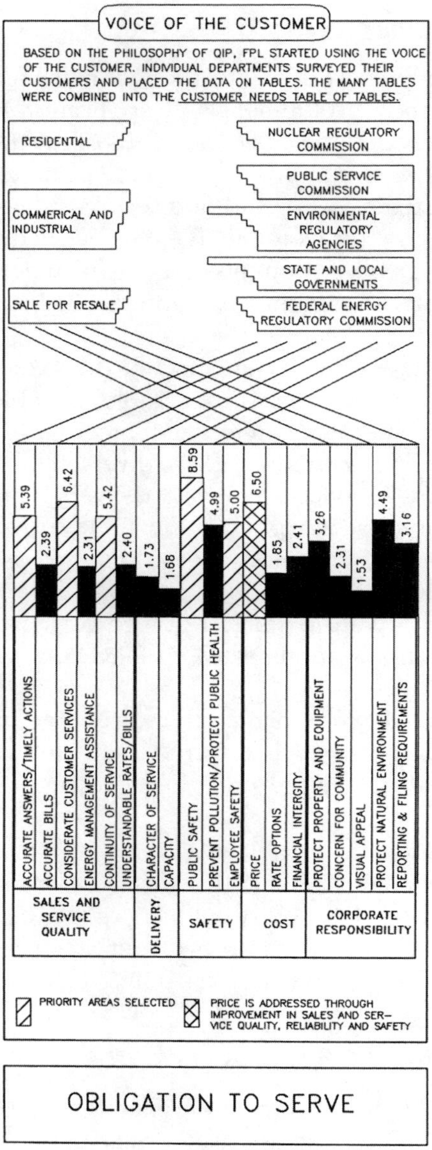

Figure 6.41 FP&L's Tables of Tables

Source: Florida Light & Power Co., *Description of Quality Improvement Program*, Corporate Unit, 1988, p. 11.

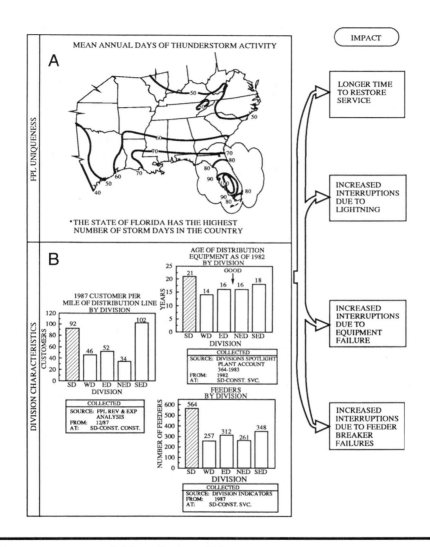

Figure 6.42 FP&L's Division Characteristics

Source: Florida Power & Light Co., *Description of Quality Improvement Program,* Southern Division, 1988, p. 2.

To counter this situation, FP&L management modified their QIP to include policy management. Policy management coordinated all divisional visions, missions, strategic plans, and business plans with each other, and with FP&L's corporate vision, mission, strategic plan, and business plan.

For example, the Southern Division's vision is as follows:* *"We will improve all aspects of operations to increase customer satisfaction and be recognized as the best managed division."* The Southern Division's vision is supported by two key objectives. They are (1) improve continuity of service as measured by "service unavailability" through the use of the QIP components, and (2) improve customer satisfaction through the use of the QIP components as measured by "complaints to the Florida Public Service Commission" and "customer satisfaction survey indicators."

The Southern Division's objectives were derived from an examination of "priority problems at the time FP&L management introduced QIP" and the "Southern Division's vision (Figure 6.43). The "priority problems at

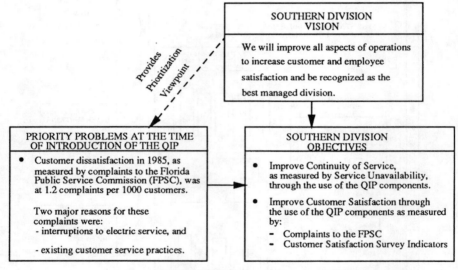

Other Objectives of Southern Division
- Continue employee safety program
- Continue cost control activities

Note: In 1991, FP&L switched from being organized by division to being organized by function. This switch was based on a scan of FP&L's environment. The scan indicated the need to streamline FP&L to make it more competititive in the 1990s.

Figure 6.43 Southern Division's Objectives

Source: Florida Power & Light Co., *Description of Quality Improvement Program,* Southern Division, 1988, p. 6.

---

* This vision seems to stimulate competition between divisions and would violate Deming's concept of globally optimizing the system of interdependent stakeholders.

the time FP&L management introduced QIP" resulted from the four crises that shaped FP&L's corporate vision, mission, strategic plan, and each division's vision. Likewise, all other divisions' vision and key objectives reflect corporate policy.

Corporate policies are established by the Executive Committee with the help of the Table of Tables and an examination of FP&L's business environment. Once set, policies are communicated to each division in the form of strategic (long-term) plans. In response, each division submits an annual business (short-term) plan detailing its predicted contribution toward the realization of the corporate strategic plan.

For example, the Table of Tables representing the "Voice of the Customer" and "problems at the time of QIP introduction" were used to guide the development and approval of business (short-term) plans as follows:

1. The Table of Tables (see Figure 6.44A) prioritized the concerns of FP&L's diverse customer base and helped FP&L's Executive Committee define the areas that were of greatest importance to customers. For example, the Table of Tables has a category titled "sales and service quality," under which fall the following related items: accurate answers/timely actions, accurate bills, considerate customer service, energy management, continuity of service, and understanding rates/bills. All these items were important to FP&L's customers. However, of these items, only "accurate answers/timely actions," "considerate customer services," and "continuity of service" were identified by the customers as high-priority items.

2. Corporate priorities were established by the Executive Committee through an examination of the problems FP&L was experiencing at the time. The Executive Committee uses the information gathered by end-of-year reviews of internal customers, and an assessment of FP&L's competitive environment to update existing strategic plans (see Figure 6.44B). These problems then are communicated to each division by the Policy Deployment Committee in the form of updated strategic (long-term) plans.

3. Each division submits annual business plans, including budgets, to the Policy Deployment Committee. Next, the Policy Deployment Committee reviews the business plans and budgets, and approves those which will have the greatest impact on achievement of the corporate strategic plan.

All business plan items (called short-term plans (STP)), (Figure 6.44C) and budget items submitted to the Policy Deployment Committee by the Southern Division were evaluated in light of the "Voice of the Customer" (Figure 6.44A) and the "problems at the time of QIP introduction" (Figure 6.44B).

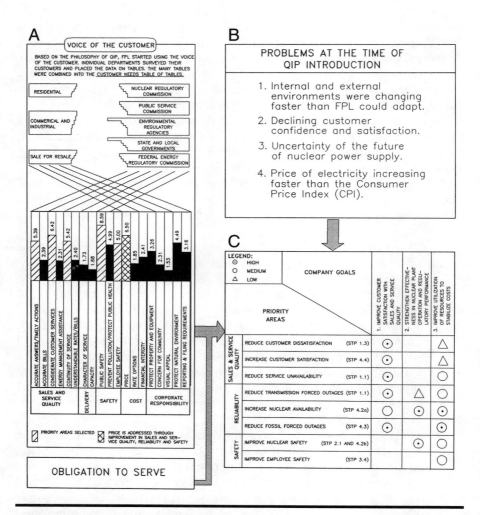

Figure 6.44 Business Plan Items

Source: Florida Power & Light Co., *Description of Quality Improvement Program,* Corporate Unit, 1988, p. 11.

In this case, the Southern Division's STP 1.1 addresses (1) the "Voice of the Customer" in respect to "sales and service quality" by focusing on "reducing service unavailability," and (2) "problems at the time of QIP introduction" by improving "reliability" (as viewed in respect to declining customer confidence and satisfaction) and "reducing transmission forced outages." Unfortunately, there is some degree of subjectivism in aligning the "Voice of the Customer" and "the problems at the time of QIP introduction" with STPs. This process is improved over time using the PDSA cycle.

4. The Policy Deployment Committee examines each of the division's STPs in light of their contribution to the corporate strategic plan (see Figure 6.44c for the Southern Division's contribution to the corporate strategic plan). For example, STP 1.1 was highly correlated (⊙) with FP&L's company goal to "improve customer satisfaction with sales and service quality," was moderately correlated (O) to "improving the "utilization of resources to stabilize costs," and was weakly correlated (Δ) to "strengthening the effectiveness in nuclear plant operations and regulatory performance." Because the Southern Division's STPs were strongly correlated with the corporate strategic plan, the Policy Deployment Committee approved funding for the Southern Division's business plan.

5. The Policy Deployment Committee integrates all the divisions' business plans into a system of business plans which mutually support the corporate strategic plan. Finally, the Policy Deployment Committee sends the integrated set of business plans to the Executive Committee.

Fulfilling Divisional Objectives

Once STP 1.1 was approved and its required funding was allocated, personnel in the Southern Division began to collect and analyze data about the factors affecting "service unavailability" (SU).

Prior to 1986, SU was measured as a percentage of service available. This method of measuring performance indicated that service was available 99.991% of the time (Figure 6.45). From FP&L's viewpoint, 99.991% reliability constituted excellent performance. However, this number did not adequately reflect the customer's true requirement, which was uninterrupted service. Therefore, a new measure, based on minutes of service outage per customer, was introduced. The new indicator portrayed SU as a number to be reduced; it quantified the quality improvement needed from the customer's viewpoint.

The Southern Division's contribution to decrease minutes of service unavailability represented 21.7% of the minutes per customer, per year, by year end 1988. The 21.7% goal is a designed goal developed by examining the capability of FP&L's existing processes, external conditions, and the "Voice of the Customer"; benchmarking other utility companies; analyzing the corporate strategic plan; and determining the resources needed to improve the SU processes.

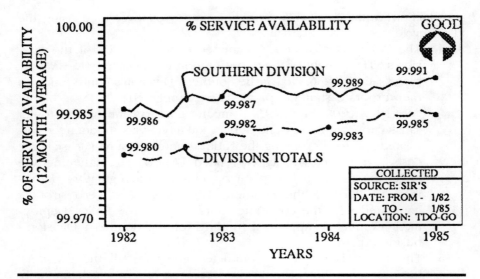

Figure 6.45 Service Availability in the Southern Division

Source: Florida Power & Light Co., *Description of Quality Improvement Program,* Southern Division, 1988, p. 46.

Once the Southern Division and the Policy Deployment Committee had agreed (via catchball) on a goal of decreasing SU by 21.7%, the Southern Division began to create a detailed plan to decrease service unavailability.

Creating a Plan

Service unavailability is measured in terms of the "frequency" and the "duration" of power outages;* Service Unavailability = Frequency × Duration. Consequently, Southern Division personnel broke down the minutes of SU per customer per year into minutes of SU per customer per year due to frequency and minutes per customer per year due to duration.

The Southern Division has certain unique characteristics that directly affect the level of service unavailability. For example, the division has a greater number of feeders,** its distribution equipment is older, and it is second only to the South-Eastern Division in the number of customers

---

* The ensuing analysis assumes that "SU" is a stable process.

** Feeders are major electric lines that carry electricity from the generating plants to a major service area or subarea. When a power outage is reported and the cause is not immediately known, work crews are dispatched to check the lines feeding the affected area.

per mile of distribution line (Figure 6.42B). These factors, along with Florida's propensity for thunderstorms and lightning strikes (e.g., Figure 6.42A), increased the frequency and duration of power outages in the area. The Southern Division's Quality Improvement (QI) team collected and analyzed data on the factors contributing to outages to better understand the causes of service unavailability.

Frequency

The Pareto chart in Figure 6.46B breaks down SU due to "frequency" (as measured by Customer Minutes Interrupted (CMI)) by major categories. Of all CMI, 56% are due to the top three categories: "Natural," "Overhead Equipment," and "Underground Equipment."

Data pertaining to these three categories, and the "Protective Device"* category, were stratified by district (Frequency Projects Assignment Matrix of Figure 6.46C). The information incorporated into this matrix was used by the QI team to assign responsibility for reducing CMI among the five districts within the Southern Division. For example, although the Hialeah district was not a leader in CMIs due to "Natural" causes, the district was already working on outages due to "Natural" causes and had made a major monetary investment, so they were assigned responsibility to reduce CMI due to "Natural" causes. On the other hand, when assigning responsibility for reductions in CMI due to "Underground Equipment," the South Dade district's large CMI due to "Underground Service" made it the obvious choice.   ·

Once they were assigned responsibility to eliminate a major category of CMI, each district in the Southern Division organized QI teams to improve relevant standardized methods.

Hialeah District Creating a Plan: "Natural" Causes

The biggest cause of CMI was attributable to "Natural" causes (Figure 6.46b). Consequently, the Hialeah District's QI team stratified "Natural" causes to gain a better understanding of its root causes (Figure 6.47a).

The Hialeah QI team identified lightning as the most significant factor affecting CMI due to "Natural" causes. Additional CMI data concerning the effects of lightning strikes on the components of the electrical distribution system (feeder, lateral, and other) were gathered and displayed on

---

* The fourth category, "Accidents," refers in part to outages caused by automobile accidents. Because many of the causes for this factor are not within FPL's control, this category was skipped in favor of "Protective Device."

- The Division Vice President's objective was to reduce Service Unavailability by 21.7% per customer by year end 1988. Policy Deployment was used to address this objective.
- In view of our characteristics we will work on Duration and four Frequency major categories.

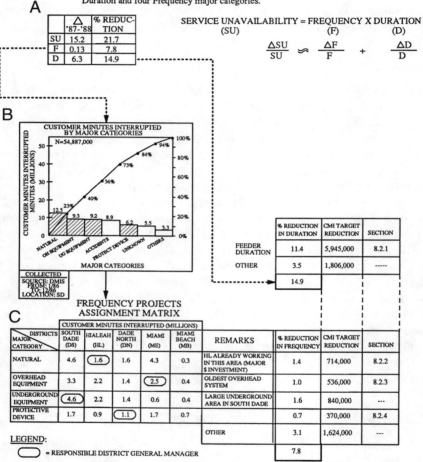

A

	△ '87-'88	% REDUC-TION
SU	15.2	21.7
F	0.13	7.8
D	6.3	14.9

SERVICE UNAVAILABILITY = FREQUENCY X DURATION
(SU) (F) (D)

$$\frac{\Delta SU}{SU} \approx \frac{\Delta F}{F} + \frac{\Delta D}{D}$$

B

CUSTOMER MINUTES INTERRUPTED BY MAJOR CATEGORIES
N=54,887,000

	% REDUCTION IN DURATION	CMI TARGET REDUCTION	SECTION
FEEDER DURATION	11.4	5,945,000	8.2.1
OTHER	3.5	1,806,000	-----
	14.9		

COLLECTED SOURCE: DMIS FROM: 1/86 TO: 12/86 LOCATION: SD

C

FREQUENCY PROJECTS ASSIGNMENT MATRIX

MAJOR CATEGORY	CUSTOMER MINUTES INTERRUPTED (MILLIONS)					REMARKS	% REDUCTION IN FREQUENCY	CMI TARGET REDUCTION	SECTION
DISTRICTS	SOUTH DADE (DS)	HIALEAH (HL)	DADE NORTH (DN)	MIAMI (ME)	MIAMI BEACH (MB)				
NATURAL	4.6	1.6	1.6	4.3	0.3	HL ALREADY WORKING IN THIS AREA (MAJOR $ INVESTMENT)	1.4	714,000	8.2.2
OVERHEAD EQUIPMENT	3.3	2.2	1.4	2.5	0.4	OLDEST OVERHEAD SYSTEM	1.0	536,000	8.2.3
UNDERGROUND EQUIPMENT	4.6	2.2	1.4	0.6	0.4	LARGE UNDERGROUND AREA IN SOUTH DADE	1.6	840,000	---
PROTECTIVE DEVICE	1.7	0.9	1.1	1.7	0.7		0.7	370,000	8.2.4
						OTHER	3.1	1,624,000	---
							7.8		

LEGEND:

⬭ = RESPONSIBLE DISTRICT GENERAL MANAGER

- Each District General Manager was made responsible for reducing CMI's (in their assigned major category) <u>throughout the entire Southern Division</u>.
- Members from the Division Engineering Department were assigned to participate in order to provide technical support.

Figure 6.46 Assignment of Responsibility for Reducing Customer Minutes Interrupted

Source: Florida Power & Light Co., *Description of Quality Improvement Program,* Southern Division, 1988, p. 47.

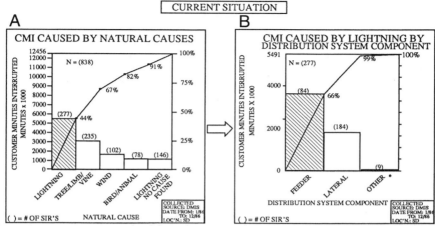

Note: In 1991, FP&L switched from being organized by division to being organized by function. This switch was based on a scan of FP&L's environment. The scan indicated the need to streamline FP&L to make it more competitive in the 1990s.

Figure 6.47 Hialeah District Creating a Plan

Source: Florida Power & Light Co., *Description of Quality Improvement Program,* Southern Division, 1988, p. 52.

a Pareto chart (Figure 6.47B). From Figure 6.47b it is clear that lightning damage to feeders accounted for 66% of CMI due to "Natural" causes. Further analysis of this problem was initiated through a cause-and-effect diagram (Figure 6.48).

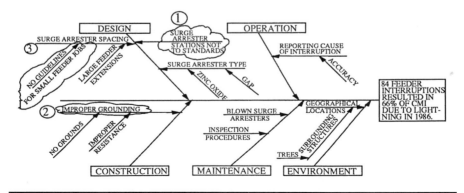

Figure 6.48 Analysis of Feeder Interrupted in the Hialeah District

Source: Florida Power & Light Co., *Description of Quality Improvement Program,* Southern Division, 1988, p. 52.

Of the possible root causes shown in Figure 6.48, "surge arrester stations not to standard," "improper grounding," and "no guidelines for small feeder jobs" were identified by the team for further analysis. From this analysis, team members identified a correlation between feeder interruptions and substandard surge arrester stations (see potential root cause 1 in Figure 6.49).

Improper grounding as a contributing factor to feeder outages was not verified as a root cause based on analysis of the data (see potential root cause 2 in Figure 6.49). Nevertheless, the QI team found that 68% of feeders were not up to FP&L standards.

Hialeah District Deploying the Plan

Immediate action was taken to bring all surge arrester stations not to standard, on feeders with one or more interruptions, up to standard.

Further preventive measures were taken to check small jobs for surge arrester protection due to the correlation between the size of the feeder job and the percentage of jobs not up to standard (see Second Stage Analysis/Countermeasure in Figure 6.49). This included using daily management as a vehicle for implementing a revised best practice method to ensure proper surge protector design for small jobs.

Hialeah District Verifying the Effectiveness of the Plan

The QI team verified the effectiveness of the plan by collecting data from Service Interruption Reports (SIRs) before and after the plan was put into effect. A comparison of the data collected on CMI interruptions due to lightning before deployment of the countermeasures and after deployment clearly indicates that the power outages due to lightning strikes were decreased significantly (Figure 6.50).

Hialeah District Standardizing Policy

These data led to standardization of the revised best practice methods for all divisions of FP&L.

Miami District Creating a Plan: Overhead Equipment

A QI team from the Miami District examined the reasons for "Overhead Equipment" failure (see Figure 6.46C) and organized the causes in a Pareto chart (Figure 6.51A). They found that the highest cause of failure was attributable to conductors.

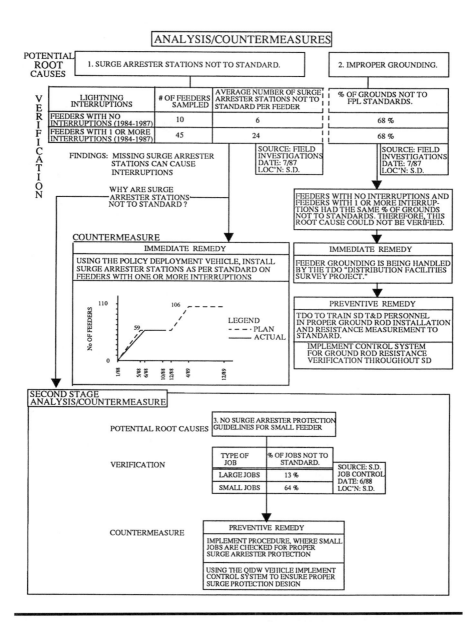

Figure 6.49 Analysis and Countermeasures in Hialeah District

Source: Florida Power & Light Co., *Description of Quality Improvement Program,* Southern Division, 1988, p. 53.

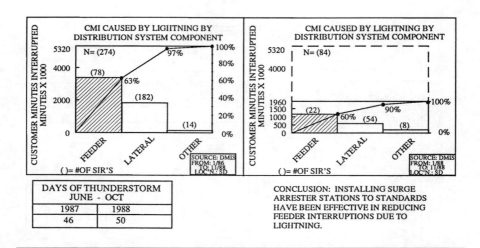

Figure 6.50

Source: Florida Power and Light Co., *Description of Quality Improvement Program,* Southern Division, 1988, p. 54.

Further analysis of conductor failures by the type of distribution system component revealed that the highest number of conductor failures occurred through the "feeder" distribution system (Figure 6.51B). Stratification of the 62 feeder data points was accomplished by asking, "What size and type of conductor is breaking?"

The answer to this question narrowed the scope of the possible root causes to #2AL conductors (Figure 6.51C). Further stratification of the data was accomplished by asking, "Is there a high failure rate with #2AL conductors? Or, are there simply a lot of #2AL conductors?" The answers to these questions are illustrated in Figure 6.51D. This table clearly shows that #2AL conductors make up 20% of the total number of conductors and account for 58% of all conductor failures. An analysis of #2AL conductors indicated that they were protected against power surges in excess of 500 Amps. However, #2AL conductors failed when power surges exceeded 200 Amps (see Figure 6.52).

Miami District Deploying the Plan

As a result of their findings, the QI team revised best practice methods to include the installation of fuse switches on all #2AL feeder branches to protect them against power surges in excess of 200 Amps. The revised best practice methods were implemented in June 1988 and were completed by December 1988.

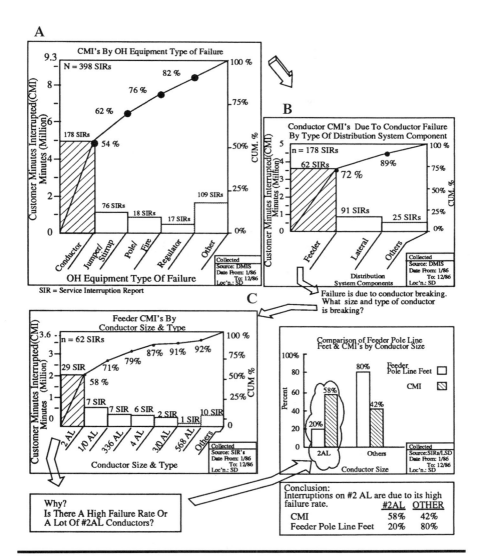

Figure 6.51 Miami District Creating a Plan

Source: Florida Power & Light Co., *Description of Quality Improvement Program,* Southern Division, 1988, p. 55.

Miami District Verifying the Effectiveness of the Plan

The QI team observed there was an initial drop in CMI due to #2AL failures by tracking the number of CMIs and their duration after the

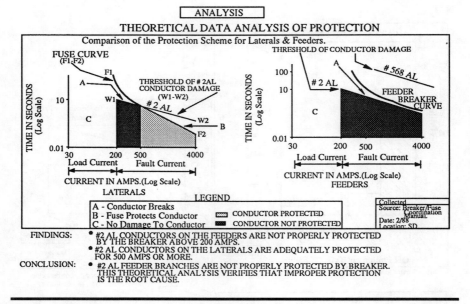

FINDINGS: • #2 AL CONDUCTORS ON THE FEEDERS ARE NOT PROPERLY PROTECTED
BY THE BREAKER ABOVE 200 AMPS.
• #2 AL CONDUCTORS ON THE LATERALS ARE ADEQUATELY PROTECTED
FOR 500 AMPS OR MORE.

CONCLUSION: • #2 AL FEEDER BRANCHES ARE NOT PROPERLY PROTECTED BY BREAKER.
THIS THEORETICAL ANALYSIS VERIFIES THAT IMPROPER PROTECTION
IS THE ROOT CAUSE.

Figure 6.52 Miami District Analyzing the Situation

Source: Florida Power & Light Co., *Description of Quality Improvement Program,* Southern Division, 1988, p. 57.

implementation of the plan. However, as illustrated in Figure 6.53, the number of interruptions continued to increase.

Miami District Standardizing the Plan

A policy statement was issued to standardize the installation of fuses on feeder branches in the Southern Division. Further, installation of fuse switches on #2AL feeders was planned for the Hialeah and South Dade districts. The QI team had uncovered one part of the problem, but as evidenced by the data, the problem of #2AL failures was not completely resolved. Continued study of the #2AL conductor failure problem was indicated to assure all possible causes and solutions had been exhausted.

Dade North District Creating the Plan: Protective Devices

A QI team from the Dade North District was established to study the causes of CMI due to "Protective Devices" (see Figure 6.46C). The team members discovered that the predominant cause of CMIs attributable to

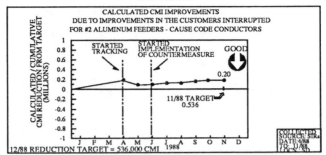

COUNTERMEASURE
INSTALL FUSE SWITCHES ON # 2 AL FEEDER BRANCHES
EFFECTS

BEGAN THE IMPLEMENTATION OF THE COUNTERMEASURE IN JUNE 1988.
WE PLAN TO COMPLETE INSTALLING FUSE SWITCHES IN MIAMI,
DADE NORTH & MIAMI BEACH DISTRICTS IN DECEMBER 1988.

STANDARDIZATION

● A POLICY STATEMENT WAS ISSUED TO STANDARDIZE THE INSTALLATION OF FUSES
ON FEEDER BRANCHES IN SOUTHERN DIVISION.

FUTURE PLANS

● INSTALL FUSE SWITCHES ON # 2 AL FEEDER BRANCHES IN HIALEAH,
AND SOUTH DADE DISTRICTS. THIS WILL COMPLETE THE INSTALLATION OF
FUSE SWITCHES.

● CONTINUE TO INVESTIGATE #2 AL CONDUCTOR FAILURES ON FEEDERS
AND LATERALS

Figure 6.53 Miami District Verifying the Effectiveness of the Plan

Source: Florida Power & Light Co., *Description of Quality Improvement Program,*
Southern Division, 1988, p. 57.

"Protective Devices" was the failure of breakers (Figure 6.54, left panel).
Further, of the breakers that failed, the ones that could not be closed
(returned to operative mode) either manually or remotely accounted for
the highest CMIs (Figure 6.54, right panel).

Stratification of the data on breakers that failed and could not be
closed (41 cases) indicated that "binding linkage" was the most significant
cause of failures (binding linkages accounted for 62.46% of CMI); see
Figure 6.55. Continued examination of the 22 failures due to binding
linkage revealed that ITE-VBK linkages and GE ML-10 linkages were the
two linkage types most likely to fail and were the two linkage types
with the greatest contribution to CMI (Figure 6.56, top panel). An
examination of the number of ITE-VBK linkages and GE ML-10 linkages
in use made it clear that ITE-VBK linkages were a more significant
contributor to CMI than GE ML-10 linkages (see Figure 6.56). The QI

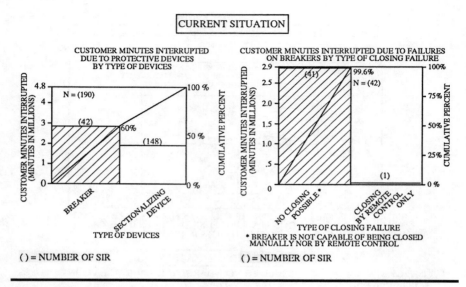

Figure 6.54 Dade North District

Source: Florida Power & Light Co., *Description of Quality Improvement Program,* Southern Division, 1988, p. 58.

team found that ITE-VBK linkages were failing at a higher frequency than other linkages because they were not within tolerances (Figure 6.56, bottom).

The team's proposed revised best practice methods provided for more specific maintenance procedures for ITE-VBK breakers and additional training in respect to maintenance procedures.

Dade North District Standardizing the Plan

The revised best practice method was very successful in reducing the ITE-VBK linkage problem (see Figure 6.57). As a result, revised best practice methods were instituted in the Dade North District, along with Industrial and Perrine Service Centers, to standardize the maintenance procedures in place prior to February 1989.

Duration

Duration represents the average number of minutes each customer is without electrical service. It is computed by taking the total number of Customer Minutes Interrupted (CMI) and dividing by the total number of

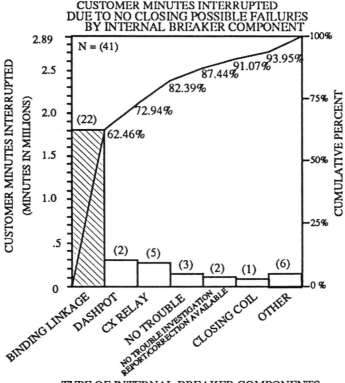

Figure 6.55 Analysis of Protective Devices in Dade North District

Source: Florida Power & Light Co., *Description of Quality Improvement Program,* Southern Division, 1988, p. 58.

customers interrupted. A QI team in the Southern Division stratified CMI duration data for a 6-month period from July to December 1986 by type of distribution system component and determined that feeders were the major cause of CMI due to duration (Figure 6.58). The duration of actual and predicted service outages due to feeders in the Southern Division was plotted over time to gain a better understanding of how feeder failures affected total CMI (Figure 6.59).

ANALYSIS

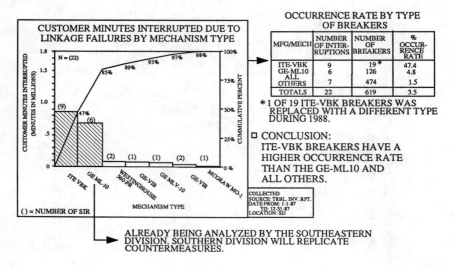

Figure 6.56 Examination of Linkages in the Dade North District

Source: Florida Power & Light Co., *Description of Quality Improvement Program,* Southern Division, 1988, p. 59.

Examining the Feeder Problem

Next the QI team compared the Southern Division's CMI due to duration against the records of all other divisions. The finding was that the Southern Division had the worst record and the Western Division laid claim to the

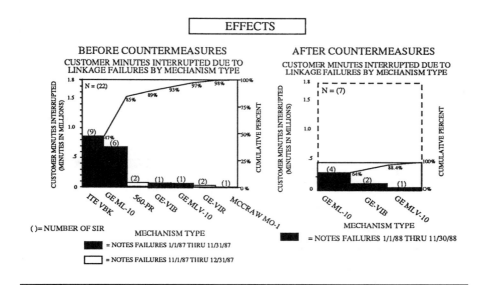

Figure 6.57 Standardizing the Plan in the Dade North District

Source: Florida Power & Light Co., *Description of Quality Improvement Program, Southern Division,* 1988, p. 60.

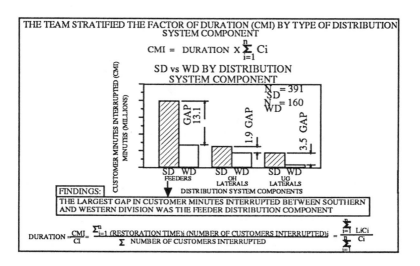

Figure 6.58 Causes of CMI Due to Duration in the Southern Division and the Western Division

Source: Florida Power & Light Co., *Description of Quality Improvement Program, Southern Division,* 1988, p. 48.

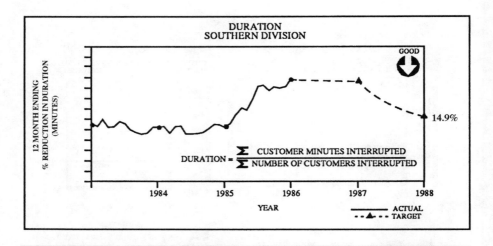

Figure 6.59 Service Outages Due to Feeders in the Southern Division

Source: Florida Power & Light Co., *Description of Quality Improvement Program,* Southern Division, 1988, p. 48.

best record, the difference being 13.1 minutes (Figure 6.58). Consequently, the Western Division was used as the benchmark for the development of "new feeder restoration" best practice method.

Flowcharts of the feeder restoration process for the Southern and Western divisions were prepared to further isolate the differences between the two divisions (Figure 6.60). The comparisons of the flowcharts revealed two key differences:

1. Sending the troubleman to designated disconnect switches called the pick-up point (PUP) (Western Division method), instead of the sub-station (Southern Division method), when the damaged location is not known, emphasizes restoration of electric service to customers first.
2. Using a binary search procedure to identify downed feeders, and thus speed up the process of isolating the damaged location (Western Division method).

In the Southern Division, when service interruptions were reported, the service dispatcher would guide the troubleman to the area where, based on the incoming calls, the damaged feeder section most likely was to be found. The troubleman would literally follow the line flow from electric pole to electric pole looking for the downed feeder. By contrast, the Western Division had established midpoints between feeders and

ANALYSIS

The team developed the feeder restoration processes for
the Western Division (best Division) and Southern Division (worst
Division) and compared them.

The comparison of processes flows revealed two differences
(Shown in Clouds below)

1. Sending the Troubleman to designated disconnect switches called the
Pick-Up-Point (PUP), instead of the substation (SD Method), when the
damaged location is not known, emphasizes restoration of electric
service to customers on first.

2. Western Division pre-determines all open points for feeders, which
eliminates the time required to select 50% point.

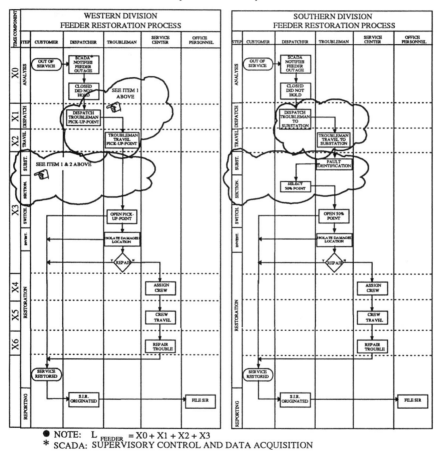

● NOTE: $L_{FEEDER} = X0 + X1 + X2 + X3$
* SCADA: SUPERVISORY CONTROL AND DATA ACQUISITION

**Figure 6.60 Flowcharts of the Feeder Restoration Process for the Southern and
Western Division**

Source: Florida Power & Light Co., *Description of Quality Improvement Program,*
Southern Division, 1988, p. 49.

substations. When a power outage was reported, the troubleman was dispatched to a midpoint. That point would be checked to determine if the problem was in the line feeding that midpoint or exiting that midpoint. Power was then restored to the operational feeder by redirecting the current. The Western Division's method restored power to the greatest number of people in the shortest possible time.

Deploying the Plan

After identifying the major differences in the feeder restoration best practice methods between the Southern and Western divisions, the following changes were made to the Southern Division's feeder restoration best practice method:

1. A feeder restoration process was developed that adapted the philosophy of the Western Division's binary search practice.
2. Predetermined pick-up points were designated for all 617 feeders in the Southern Division.

These measures isolated the damaged feeder line section quickly and emphasized restoration of service by rerouting power to as many customers as possible. When a service outage report is received, the midpoint between the service outage and the substation is calculated and a troubleman is dispatched. Upon location of the problem area, service is rerouted through other feeders, or through another substation.

Verifying the Effectiveness of the Plan

The effects of the Southern Division's new feeder restoration best practice method resulted in a 5.7 minute decrease in feeder restoration time (Figure 6.61).

Standardizing Policy

The revised best practice method for feeder restoration resulted in surpassing the designed goals for CMI due to duration in 1987 and 1988 (see Figure 6.62).

Review of FP&L Case Study

FP&L's QIP is driven by the "Voice of the Customer." FP&L conducts extensive customer surveys to identify the needs of its diverse customer

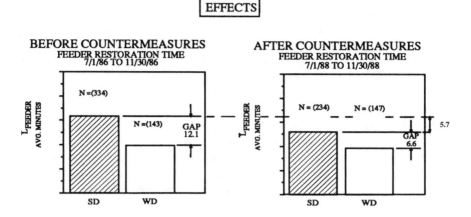

Figure 6.61 Verifying the Effectiveness of the Plan in the Southern Division

Source: Florida Power & Light Co., *Description of Quality Improvement Program,* Southern Division, 1988, p. 50

Figure 6.62 Standardizing the Policy in the Southern Division

Source: Florida Power & Light Co., *Description of Quality Improvement Program,* Southern Division, 1988, p. 50.

segments. Then, FP&L prioritizes the needs of its customers through the development and use of the "Table of Tables."

Next, the Executive Committee develops a strategic plan that reflects the needs of the customers, as well as the realities of FP&L's competitive environment. The Voice of the Customer, as reflected in the strategic plan, is disseminated throughout the organization by the Policy Deployment Committee. Next, each division submits a business plan to the Policy Deployment Committee. The business plans are approved and funded by the Policy Deployment Committee. FP&L's QIP involves everyone in the organization in setting and meeting business plans through daily management. Daily management provides a system for translating corporate policy into daily best practice methods, which can be standardized and improved via the SDSA and PDCA cycles.

Employees of FP&L were able to resolve their four crises through their QIP. The extraordinary accomplishments of FP&L employees culminated in 1989 when John J. Hudiburg, then CEO of FP&L, accepted the coveted Deming Prize from the Japanese Union of Scientists and Engineers.

FP&L IN THE 21ST CENTURY

An update on the status of FP&L's quality process in the 21st century is provided in Appendix 6D.

SUMMARY

Chapter 6 discusses policy management, Prong Three of the Quality Management model presented in this book. Policy management is performed by turning the PDSA cycle to improve and innovate the methods responsible for the difference between corporate results and corporate targets, or to change the direction of an organization. Policy management includes setting policy, deploying policy, studying policy, providing feedback to employees on policy, and conducting Presidential Reviews of policy. Policy management is accomplished through the workings of an interlocking system of committees, including the Executive Committee (EC), the Policy Deployment Committee (PDC), Local Steering Teams (LSTs), and Project Teams.

The President conducts an initial Presidential Review to determine the state of the organization and to develop a plan of action for the promotion of corporate policy. This promotes a dialogue between the President and mid-level management and brings out information about problems. After a few rounds of Presidential Reviews, the President will have a good understanding of the major problems facing the organization and their possible causes.

The EC is responsible for setting the strategic plan for the entire organization. That includes establishing values and beliefs, developing statements of vision and mission, and preparing a draft set of strategic objectives. Techniques such as an affinity diagram, S.W.O.T. analysis, and the Table of Tables are used to gather information and develop strategic objectives.

Members of the PDC develop a set of integrated improvement plans to promote the strategic objectives. The members of the PDC use tools such as gap analysis, Pareto diagrams, and cause-and-effect diagrams to develop the departmental and corporate improvement plans needed to promote departmental and corporate strategic objectives. The members of the LSTs are responsible for coordinating and carrying out the projects set up in the corporate and departmental improvement plans.

Strategic objectives and improvement plans are deployed by the PDC through assignment of responsibility for action to people or groups of people in departments. Techniques used in policy deployment include catchball and flag diagrams.

Policy is implemented when teams work on projects to improve and/or innovate processes. It is also implemented when departments use the revised processes and measure their results in respect to improvement plans and strategic objectives.

Periodic management reviews are conducted at two levels. First, the members of the EC review progress toward each strategic objective and its improvement plans monthly. Second, the members of the PDC and appropriate LSTs review progress for each project. The purpose of these reviews is to provide feedback to project team members that promotes process improvement efforts.

An application of policy management to a person's life and to Florida Power & Light Co. is presented in this chapter. These examples demonstrate all the steps of policy management and how to implement them.

Appendix 6A

THE VOICE OF THE CUSTOMER

The term "customer" includes external customers and indirect customers.* External customers are the organizations or individuals who buy or use an organization's goods or services. Indirect customers are organizations that guard the welfare of external customers; for example, regulatory commissions and governmental agencies.

The "Voice of the Customer"** is a tool used to (1) define the ever-changing market segments for customers; (2) determine and prioritize the customer requirements of each market segment; (3) identify the processes (methods) used to respond to the customer requirements of each market segment; (4) construct a matrix that explains the relationships between "the customer requirements of each market segment" and "the processes (methods) used to respond to the customer requirements"; and (5) prioritize the processes (methods) used to respond to customer requirements. Data collected from a Voice of the Customer analysis are used to formulate the strategic objectives of an organization. The steps for conducting a Voice of the Customer study are shown below.

---

* The term "indirect customers" was developed by Florida Power & Light Co. See *FPL's Total Quality Management,* Unit 12, page 11.
** The author believes that the term "Voice of the Customer" was developed by Florida Power & Light Co. Much of the information in this section is paraphrased from *FP&L's Total Quality Management — Participant Handbook,* University of Miami Institute for the Study of Quality in Manufacturing and Service and Qualtec, Inc. (an FP&L subsidiary), 1990, see Unit 12, pp. 4, 10–22.

STEP 1

Define the ever-changing market segments for customers. The term "market segment" explains the dynamic and changing homogeneous groupings of customers in respect to the demographic, psychographic, and purchasing behavior variables that affect their decision to purchase and/or use a good or service. Focus groups and surveys, as well as other tools, are used to identify customer requirements for each market segment. Special care is taken to identify and define the customer requirements of non-customers and future market segments.

STEP 2

Determine and prioritize the customer requirements of each market segment. Management collects and analyzes observational, survey, and experimental data to understand the Voice of the Customer by market segment. The question asked of a sample of customers from each market segment is, "From your perspective, what requirements must the organization surpass to pursue the mission statement?"

For example, Figure A.1* shows a prioritized list of customer requirements that were collected by randomly sampling Florida Power & Light residential customers. For each market segment, each customer requirements is scored on three scales: first, an "importance to the customer" scale; second, a "current level of performance in the eyes of the customer" scale; and third, a "desired level of performance by management to optimize the interdependent system of stakeholders" scale. The "total weight" is computed for each customer requirement in each market segment. "Total weight" is a measure of the need to take action on a customer requirement.** See the right side of Figure A.1.

The "importance to the customer" scale quantifies the importance of customer requirements for each market segment. It does not quantify how well the organization is currently handling customer requirements or how much improvement is required in respect to customer requirements. The scale is a 1 to 5 scale: 1 = very unimportant and 5 = very important. "Importance to the customer" scores are obtained by computing the average ratings for each customer requirement for each market segment from survey and/or focus group data.

---

* This figure is taken from *FP&L's Total Quality Management*, Unit 12, p. 13.
** This procedure for prioritizing customer requirements is adapted from the Quality Function Deployment methods of Dr. Akao.

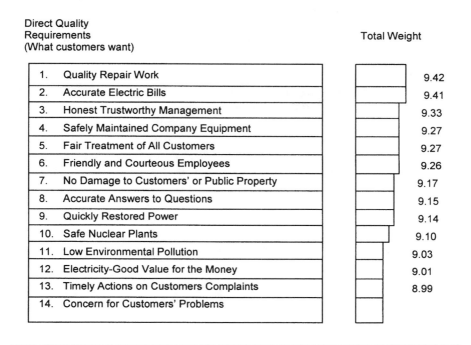

Direct Quality Requirements (What customers want)	Total Weight
1. Quality Repair Work	9.42
2. Accurate Electric Bills	9.41
3. Honest Trustworthy Management	9.33
4. Safely Maintained Company Equipment	9.27
5. Fair Treatment of All Customers	9.27
6. Friendly and Courteous Employees	9.26
7. No Damage to Customers' or Public Property	9.17
8. Accurate Answers to Questions	9.15
9. Quickly Restored Power	9.14
10. Safe Nuclear Plants	9.10
11. Low Environmental Pollution	9.03
12. Electricity-Good Value for the Money	9.01
13. Timely Actions on Customers Complaints	8.99
14. Concern for Customers' Problems	

Figure 6A.1 Prioritized List of Residential Customer's Requirements Florida Power & Light Co.

Source: FPL TQM, Qualtec, Inc., 1990.

The "current level of performance" scale quantifies the gap between customer requirements and organizational performance. This scale is a 1 to 5 scale: 1 = very large gap and 5 = very small gap. The measure of current performance scale is obtained by computing the average ratings from survey or focus group data for each customer requirement for each market segment from the following question: "How is the organization doing in respect to exceeding customer requirement x?"

Customer requirements that show a "significant" gap are targeted for further study via gap analysis. For each customer requirement and market segment, gap analysis requires a measure of customer requirements (see "importance to the customer" scale) and a measure of current organizational performance (see "current level of importance" scale) to highlight the customer requirements that should be studied further with gap analysis.

Gap analysis is a procedure for studying the root cause(s) of the difference between customer requirements and organizational performance. It is based on the analysis of relevant data. Many different tools are helpful in gap analysis; for example, flowcharting, the seven basic QC tools, and benchmarking. For example, the members of the EC might assign a group of staff personnel to study the root causes of the gap for a particular market segment. The group might study the gap over time and determine that it is stable and contains only common variation. Next, they could construct a Pareto diagram of the common causes of the gap, isolate the most significant common cause, and develop a cause-and-effect diagram of its causes. Next, the staff personnel would study the correlation between the suspected root cause and the most significant cause of the gap. If the staff personnel found the correlation to be significant, they would recommend to the members of the EC a plan of action for improving the current level of performance for that customer requirement.

The "desired level of performance" scale quantifies the desired level of performance for each customer requirement, for each market segment. The scale is a 1 to 5 scale: 1 = small improvement in the organization's ability to exceed a customer requirement, and 5 = large improvement in the organization's ability to exceed a customer requirement. "Desired level of performance" scores are developed by staff personnel assigned by the EC. They conduct analyses of the levels of performance required for each customer requirement, for each market segment, to stay ahead of other organizations in the industry (using benchmarking) and future customer requirements.

The "total weight" score is a measure of the importance of the gap for each customer requirement, in each market segment. "Total weight" scores are computed using the following formula:

$$\text{Total Weight} = [I \times D]/C$$

Where:
I = importance to the customer
D = desired level of performance
C = current level of performance

STEP 3

Identify the processes (methods) used to respond to the needs and wants of each market segment and indirect customer. It is critical that the needs and wants of each appropriate market segment or indirect customer be serviced by identifiable methods. Customer needs and wants are translated into improved and innovated methods. This translation is accomplished

Translating Customer's Requirements Into Methods

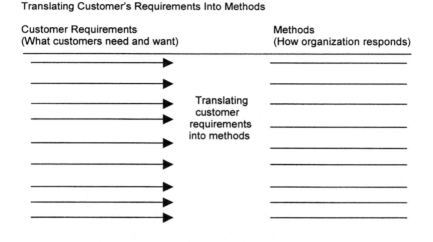

| Customer Requirements (What customers need and want) | | Methods (How organization responds) |

Figure 6A.2 Processes Necessary to Respond to Customer's Requirements

by asking what methods are necessary to respond to each customer need or want. Figure A.2 may be helpful.*

STEP 4

Construct a matrix that explains the relationships (cells of the matrix) between "customer requirements" (rows) and "the processes (methods) used to respond to customer requirements" (columns), for each market segment and each indirect customer. The matrices for all market segments and indirect customers should have the same columns. The relationships shown in the cells of each matrix are determined by a group of staff personnel assigned this responsibility by the EC. The staff uses its knowledge of the organization and customers, along with that of other knowledgeable people, to determine the relationships. The staff members assigned to determine the relationships between the rows and columns keep a record of their logic for each symbol placed in the matrix, so they will not be second-guessed at a later date. Relationships are measured on the following scale: 3 = strong relationship, 2 = moderate relationship, 1 = weak relationship, and blank = no relationship. Sometimes, a doughnut symbol is used for a 3, a circle symbol is used for a 2, and a triangle symbol is used for a 1. The above numbers or symbols are used whether

---

* This chart is paraphrased from *FP&L's Total Quality Management*, unit 12, p. 15.

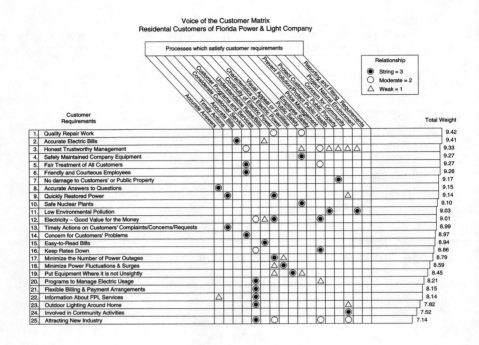

Figure 6A.3 Voice of the Customer Matrix Residential Customers of Florida Power & Light Co.

Source: Florida Power & Light Co., *Customer Needs Table of Tables*, Research, Economics, and Forecasting Department, 1990.

the relationships are positive or negative. A matrix showing the needs and wants of a particular market segment and the methods needed to respond to the needs and wants of the customers in the market segment is shown in Figure 6A.3.

Every customer requirement must be adequately serviced by one or more methods. If a customer requirement is not being serviced by any method (or not adequately serviced), one or more methods are developed to service the customer requirement. If a method is not servicing (directly or indirectly) at least one customer requirement, the method is dropped or receives decreased attention by stakeholders.

STEP 5

Prioritize the processes (methods) used to respond to customer requirements for attention in the strategic objectives of the organization.

1. Compute the unnormalized weights (see Figure 6A.4) for each process, for a given market segment or indirect customer. For a given process (column in Figure 6A.3), multiply the "total weight" score for each customer requirement by the relationship score between that process and each customer requirements, and add all products in a column. For example, the unnormalized weights for the process's "accurate answers" and "timely actions" are computed as follows:

   Accurate answers:

   $$35.59 = (9.15[3] + 8.14\,[1]),$$

where
   9.15   = the total weight for "accurate answers to questions,"
   8.14   = the total weight for "information about FP&L services,"
   3   = the relationship between "accurate answers" and "accurate answers to question,"
   1   = the relationship between "accurate answers" and "information about FP&L services."

   Timely answers:

   $$54.39 = (9.14[3] + 8.99\,[3]).$$

See Figure A.4.
2. Normalize the weighted values by dividing the individual weighted values by the sum of all weighted values (see Figure A.4).
3. Prioritize the normalized weighted values over all methods to provide input into the selection of strategic objectives for the organization. For example, the above analysis indicates that "timely answers" would receive a higher priority for attention than "accurate answers."

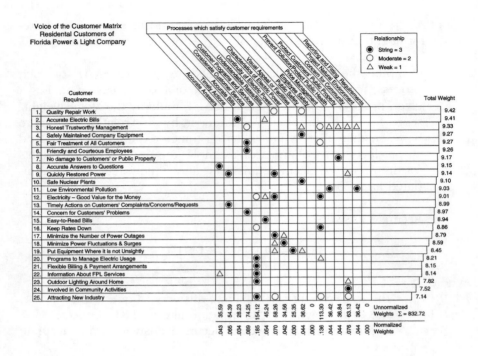

Figure 6A.4 Voice of the Cutomer Matrix Residential Customers of Florida Power & Light Co. with Normalized Weights

Source: Modified from Florida Power & Light Co., *Customer Needs Table of Tables,* Research, Economics, and Forecasting Department, 1990.

Appendix 6B

THE VOICE OF THE BUSINESS

The "Voice of the Business" is a tool for collecting and analyzing data about employee requirements (e.g., concerns and fears) in respect to the mission of an organization. Voice of the Business studies require that all groups of employees answer the following question: "What requirements (e.g., concerns and fears) do you have in respect to the organization pursuing its mission?" Data from the answer to this question helps to formulate the strategic objectives of an organization. The procedure for conducting a "Voice of the Business" study is described below.

STEP 1

Collect and analyze the answers to the question posed above for each employee group, for example, top management, middle management, first line supervisors, and hourly employees. Brainstorming sessions, focus groups, surveys, and management reviews are examples of tools that are useful in collecting information about the above question. Affinity diagrams, interrelationship diagraphs, and cause-and-effect diagrams are examples of tools that can be used to analyze the answers to the above question.*

STEP 2

Determine and prioritize the employee requirements of each employee segment, see step 2 in Appendix 6A. Employee requirements are determined for each employee group, just as customer requirements are determined for each market segment and indirect customer.

---

* These tools are discussed in Gitlow, H. and PMI, *Planning for Quality, Productivity and Competitive Position*, Dow Jones-Irwin (Homewood, IL), 1990.

STEP 3

Identify the processes (methods) used to surpass the employee requirements for each employee group. These processes are the same as or additions to the processes used to address the customer requirements in a Voice of the Customer study. Employee requirements are used in developing strategic objectives for the organization.

STEP 4

Construct a matrix that explains the relationships (cells of the matrix) between "employee requirements" (rows) and "the processes (methods) used to respond to employee requirements" (columns), for each employee segment. All Voice of the Customer and Voice of the Business matrices should have the same columns. See step 4 in Appendix 6A.

STEP 5

Prioritize the processes (methods) used to respond to employee requirements for attention in the strategic objectives of the organization. See step 5 in Appendix 6A.

Appendix 6C

TABLE OF TABLES

The original concept of the "Table of Tables" was developed by the members of the Research, Economics, and Forecasting Department of Florida Power & Light Co. The Table of Tables presented here is a variant of the Florida Power & Light Table of Tables. It considers customer requirements and employee requirements. The Table of Tables creates one prioritized list of processes (methods) to be highlighted for attention in the strategic objectives of the organization.

BUILDING A TABLE OF TABLES*

A graphic mock-up of a Table of Tables can be seen in Figure 6C.1. Please note that all the Voice of the Customer subtables on the left side of the Table of Tables (there are three) and the Voice of the Business subtables shown on the right side of the Table of Tables (there are five) share a common set of processes (methods) in their columns. The Voice of the Customer and Voice of the Business studies all result in prioritized lists of the common set of processes (methods). The Table of Tables globally prioritizes the common processes from each subtable.

The group of staff employees selected by the EC establishes a weight for the process priorities from each direct and indirect customer group (subtable) in the Voice of the Customer analysis and a similar weight for each employee group in the Voice of the Business analysis. For example, each of the three customer groups on the left side of Figure 6C.1 could receive equal weights of 0.333, 0.333, and 0.333, while the five employee groups on the right side of Figure 6C.1 could receive weights determined

---

* Modified from Research, Economics, and Forecasting Department, *Supplement to the Customer Needs Table of Tables*, version 5, Florida Power & Light Company, 1990, p. 3.

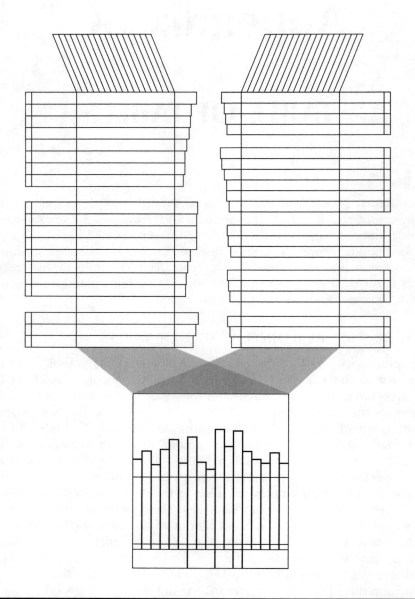

Figure 6C.1 Generic Table of Tables

Source: Florida Power & Light Co., *Customer Needs Table of Tables*, Research, Economics, and Forecasting Department, 1990.

by the members of the EC, which are based on their assessment of the impact of each employee group on the organization. For example, employee group 1 may receive 0.10, employee group 2 may receive 0.50, employee group 3 may receive 0.05, employee group 4 may receive 0.05, and employee group 5 may receive 0.30.

The normalized scores in each subtable are multiplied by the appropriate subtable weight. The weighted normalized scores are summed over all tables for the left and right sides of the table of tables.

The members of EC weight the relative impact of the Voice of the Customer and the Voice of the Business on the organization. For example, the Voice of the Customer weighted normalized scores may receive a relative importance of 0.75, while the Voice of the Business normalized scores receive a relative importance of 0.25.

Finally, the weighted normalized scores for the Voice of the Customer and the Voice of the Business are multiplied by the relative weights of each voice. This results in one prioritized list of processes (methods) to be used as input in the selection of strategic objectives for the organization.

CASE STUDY

The Table of Tables developed by Florida Power & Light Co. in 1988 to input the Voice of the Customer data into their policy management process is shown in Figure 6C.2. Figure 6C.2 demonstrates how FP&L defined "direct needs" as being the desires of their three market segments for external customers (residential, industrial, sale for resale); only residential and sale for resale are shown. Figure 6C.2 also shows how FP&L defined "indirect needs" as being the concerns of their regulators and government agencies (NRC, PSC, ERA, state and local governments, and the FERC); only NRC, PSC, and FERC are shown. Additionally, Figure 6C.2 shows the pie charts containing the weights for the "direct needs" and the weights for the "indirect needs." These weights are multiplied by each normalized weight in a column from the table of tables. For example, for direct needs, 0.72 is multiplied by the normalized weight for "Accurate Bills" (0.065) in the residential direct needs market segment, 0.27 is multiplied by the normalized weight for "Accurate Bills" in the commercial/industrial direct needs market segment, and 0.01 is multiplied by the normalized weight for "Accurate Bills" in the resale direct needs market segment. Please note that 0.72 + 0.27 + 0.01 = 1.00. Finally, the weighted averages for the direct needs and the weighted averages for the indirect needs, for each column, in the Table of Tables are averaged. This yields an overall priority values for each process (column) in the Table of Tables. These priority values indicate which processes (columns), when given attention for improvement, will maximally affect the most stakeholders. The Table of Tables demonstrates how FP&L prioritized its quality elements (what it does, or processes) to better satisfy customer needs and wants via policy management.

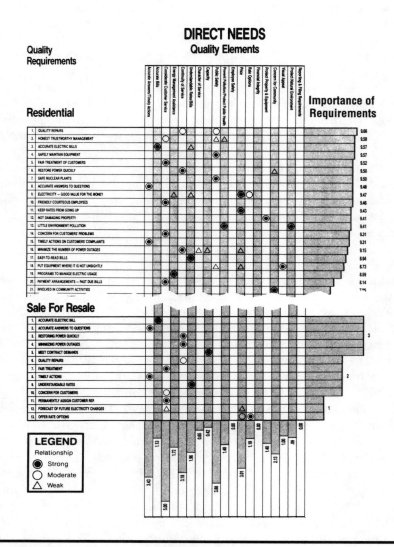

Figure 6C.2 Customer Needs Table of Tables

Source: Florida Power & Light Co., *Customer Needs Table of Tables*, Research, Economics, and Forecasting Department, 1990.

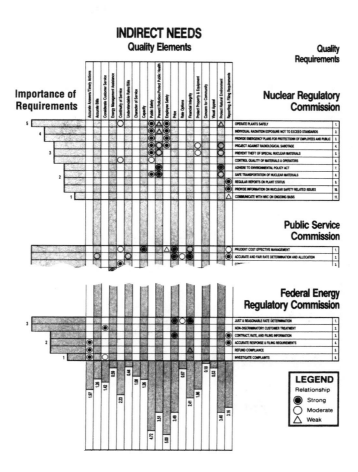

Figure 6C.2 (continued) Customer Needs Table of Tables

Customer Needs Table of Tables

**Weighting of
Direct Needs**

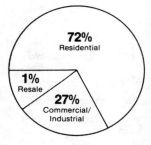

**Weighting of
Indirect Needs**

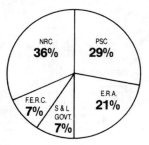

NOTES
*Importance of Requirements

1. Residential and C/I: Mean rating (scale of 1-10) of customers responding to Satisfaction Survey.

2. Resale: 1-3 rating based on judgement of staff representing customer segment.

3. NRC, ERA: 1-5 scale based on consequence of violation.

4. Gov't, PSC, and FERC: 1-3 scale based on consequence of violation.

The Table of Tables represents FPL's customers' needs and their importance ratings of these needs. It does not represent the company's ranking of functional areas.

FPL

©Copyright 1990 Florida Power & Light Company
Research, Economics and Forecasting Department

Overall Ranking of Corporate Quality Elements

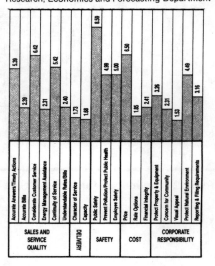

Figure 6C.2 (continued) Customer Needs Table of Tables

Appendix 6D

THE EVOLUTION OF QUALITY AT FLORIDA POWER & LIGHT*

The quality process at Florida Power & Light Company (FP&L) has evolved and matured through multiple phases, those leading up to the Deming Prize in 1989 and, equally important, those that evolved or advanced the process to respond to market changes. The four phases defining the evolution of quality are the preliminary phase, intensification phase, mature phase, and the high performance phase.

PRELIMINARY PHASE (1981–1985)

The first formal quality efforts began in 1981 with the introduction of quality improvement teams. Top management was encouraged to learn more about the TQM process as a means to address current challenges of the company. Operating costs were rising higher than inflation and customer dissatisfaction was increasing. Despite being a natural monopoly, management believed there was a better way to run the business. After reviewing several models, FPL selected the Japanese model and engaged the services from the Union of Japanese Scientists and Engineers (JUSE) and established a relationship with The Kansai Electric Power Company in Osaka, a 1984 Deming Prize winner. With the assistance of its Japanese friends and counselors, the quality improvement program (QIP) was developed and implemented company-wide in 1985. The

* This appendix was prepared by J. Michael Adams, Director of Quality, FP&L Company/FPL Group, Inc., May 2000. The author wishes to express thanks to J. Michael Adams and FP&L for their generous support in the preparation of this case study.

employees during this phase were introduced to the seven basic quality improvement (QI) tools and team management techniques.

INTENSIFICATION PHASE (1986–1989)

The initial phase was succeeded by the intensification phase that brought a stronger alignment to company performance and further developed capabilities of the FP&L's employees. To inculcate the skills into the culture, employees were required to use a rigorous seven-step problem solving process, called the "QI Story," when attempting to solve problems. Employees were reviewed for their abilities in demonstrating their use of the seven basic QI tools. In this phase, two new elements were added to evolve into a TQM system. These were Policy Deployment and Quality in Daily Work. Policy Deployment enabled employees to focus on a few company-wide high priority issues rather than dilute their impact by working on local matters of lesser impact. Quality in Daily Work provided the beginning of statistical process control and strengthened the understanding and relationship of internal customers to external customers. Two percent of the employees were trained as "Application Experts" (AEs) in more sophisticated tools and techniques. They learned and applied tools and techniques such as Weibull Analysis, Failure Modes and Effects Analysis, Regression Analysis, Design of Experiments, and general reliability tools and techniques.

The three elements of Teams, Policy Deployment, and Quality in Daily Work constituted a TQM quality system supported with an employee education and development foundation. Two structures were launched to reinforce the new culture, application of tools and techniques, and sharing solutions. The EXPO structure was a convention-like atmosphere with booths, exhibits, or storyboards displaying various improvements. In 1987, the President's Cup Team Competition began as a mechanism to judge the "quality of quality" throughout the organization and role model various tools and techniques reflective of the development strategy.

Around the same time, hundreds of visitors were visiting FP&L to gain further insight into the quality movement. Although FP&L had adopted the Japanese quality approach, it played a leadership role in the founding of the Malcolm Baldrige National Quality Award, testifying to the U.S. Congress as to its benefits for the American economy, crafting the criteria, and assisting with endowing the award.

In summer of 1988 a decision was reached that FP&L would challenge for the Deming Prize. The challenge that lasted for almost an entire year required employees to accelerate the pace of improvements and to put in many extra hours being further developed or preparing for the exam itself. Commensurate with the effort, building TQM capabilities also provided operational benefits, including substantial reduction in unplanned power plant outages and dramatic reductions in customer complaints, service reliability, and personnel injuries. In November 1989, after an intensive 2-week on-site examination, FPL became the first non-Japanese company to win the Deming Prize.

THE 1990S — AFTER THE DEMING PRIZE

The Deming Prize is presented to an organization that demonstrates the potential to systematically improve its performance. A quality system therefore must endure and accommodate change to be successful in its performance objectives.

The early 1990s posed a series of changes internally and externally to FP&L including a new CEO, a changing marketplace, and the catastrophic Hurricane Andrew, all of which had impact on the approach, deployment, and reinforcement of the tenets of quality.

MATURE PHASE (1990–1997)

The new CEO, James L. Broadhead, recognized the changing marketplace as well as the potential of the organization. He altered the course of FPL to compete in a deregulated marketplace. A key success factor in a competitive marketplace is to be a low cost provider. In 1989 however, FPL was the highest cost major electric utility in its region.

While winning the Deming Prize was an honor of which all employees were proud, there was a widespread feeling among employees that FP&L's quality program had become very mechanical and inflexible. The paper-oriented bureaucracy that served the organization in becoming developed was actually now creating barriers to continuous improvement. At this point in the quality journey, the employees were a well-developed, homogenous group with thorough process and tools knowledge. At the same time, the rate of change toward deregulation was underestimated and the transition from monopoly to competition would be realized sooner, thus requiring a more flexible, faster organization. The mature phase is one that fully integrated the TQM components into the general business structure.

An employee team was assembled to review and provide recommendations to advance the quality program. Some of the recommendations included easing up on the mandatory structures that prohibited the benefits of the workforce's knowledge to be applied toward performance rather than process. Some primary changes to the system included no longer requiring the seven-step QI story for all problem solving, ultimately dispersing the rather large quality staff to positions within the operating organization, and maintaining a small corporate quality office as part of a company-wide reorganization. This office would interface with contemporary marketplace practices like benchmarking and reengineering to keep the organization current. Quality was positioned as part of the way FP&L does business. Quality plans and business plans were no longer separate and management would take on the role of facilitator and coach to improvement as a part of their job.

The reorganization was the outcome of eight employee teams researching various disciplines including human resources, the marketplace, regulatory issues, and technological advancements. With it came a new vision: "to be the preferred provider of safe, reliable, cost effective electricity-related products and services for all customer segments."

Supporting the vision included four areas of focus listed as strong customer orientation, commitment to quality, cost-effective operations, and speed and flexibility that were maintained throughout the 1990s, adding safety in 1999. The Divisions and District geographic areas and layers of management were eliminated, reducing the layers from 12 to 5. In light of the changing marketplace, and the new vision and areas of focus, the performance measures were also changed commensurate with a competitive market. With the new direction, the capabilities could be exercised to perform with alignment and contribution to the areas of focus.

In 1994 FPL underwent a Post-Deming Prize Review at the request of the Chairman. Oversees winners of the Deming Prize can volunteer for the review where in Japan it is mandatory. It is usually conducted 3 years after the award. FPL's original request was deferred, however, to allow for the restoration and recovery from Hurricane Andrew in August 1992. Dr. Kume and Dr. Akao conducted the review. Both were Deming Prize examiners in 1989 and requested to review those sites they had previously visited and knew thoroughly.

They reviewed power plants and a customer care center, and conducted sessions with most operating and staff groups, concluding with the executive team. They appraised the organization's realization of its potential for business performance, citing best practices in benchmarking, empowerment, creative use of technology, and quality promotion. As with any review, they also provided guidance to further improve the system. Jim Broadhead was asked to present FPL's evolution and results of the review at the International Conference on Quality in Yokohama, Japan in 1996.

HIGH PERFORMANCE PHASE (1997–PRESENT)

In 1997, the quality system evolved to further advance the organization commensurate to the competitive marketplace while assessing the new fledgling business initiatives at FP&L and the skills of a new, growing workforce. Using a customized version of the Baldrige Criteria, a team of internal experienced assessors performed assessments on business units and gauged the organization relative to the four phases. Sample characteristics describing a high performance organization include work systems for enterprise profitability, leveraged use of technology, backward and forward integration, customization, upper decile performance, and leveraged learning. In 1999, following nine business unit assessments, a core team of examiners analyzed its findings and drafted recommendations for organizational continuous improvement to the senior executives.

FPL concluded the decade and millennium with the high levels of performance as well as continued contribution and leadership to the quality movement. The EXPO continues as a sharing opportunity for all employees of FPL Group, Inc. while the President's Cup Team competition enters its 14th year. The 1990s proved successful with dramatic improvements. Decade-end performance was at all-time highs with segmented customer satisfaction levels at targeted levels. Fossil plant availability reached 93% (up from 77% in 1990), while nuclear availability reached 94% (up from 77% in 1990). The power plants performed at upper decile performance levels and have provided that capability to FPL Energy, a new power generation subsidiary of FPL Group, Inc. Other performance achievements include O & M cost reduction of 36% since 1990 at residential electricity prices 16% below those of 1985 (8.34 cents/kWh compared to 6.97 cents/kWh). Considering inflation, that equates to over a 60% difference. Service reliability had some variation in the early decade but since 1997 alone, there has been a 45% improvement in reliability at levels significantly better than the national average. Improvement occurred in reducing the average length of interruption as well as the frequency.

Jim Broadhead was 1999–2000 President of the Foundation for the Malcolm Baldrige National Quality Award. Other employees serve as judges and examiners for various quality awards in the nation.

The quality system continues to evolve, driven by market condition and practices and certainly with lessons learned throughout.

7

RESOURCE REQUIREMENTS OF THE FORK MODEL

INTRODUCTION

At this point in your study of the fork model, you are probably asking yourself some pretty important questions, such as: "How long will it take my organization to 'live' the fork model? How much will it cost? What resources will I need from my organization? What resources will I need from a consulting firm?" The purpose of this chapter is to give you a template (see Template) for answering some of your questions. It is important that you realize that this chapter only presents a sample template. You will have to modify it for your organization.

There will be tremendous variation between organizations in their answers for each of the above questions. Your answers will provide an estimate of the resources necessary for the transformation of your organization.

THE TEMPLATE

Generally, the fork model is implemented sequentially; that is, the handle comes first, then the neck, then prong 1, then prong 2, and finally, prong 3. Consider each part of the fork model as a phase of the Quality Management implementation process. Different implementation strategies may be used based on the characteristics and needs of your organization.

The following abbreviations are used throughout the template.

P President
EC Executive Committee

LC   Lead Consultant
PDC Policy Deployment Committee
PIL   Process Improvement Leader
PITM Process Improvement Team Member
LST Local Steering Team
CFPL Cross Functional Project Leader
CFTM Cross Functional Team Member

DISCUSSION OF THE TEMPLATE

The template shown on the following pages is a tool that can help top management answer some of their questions about Quality Management. The template provides rough estimates for the time required to initially promote Quality Management in an organization in which top management is seriously committed to Quality Management. The model shows a minimum of 8 months to determine management's commitment to transformation, a minimum of 4 months to affect management's values and beliefs about business through education, a minimum of 4 months to produce results from daily management, a minimum of 6 months to begin cross-functional management, and a minimum of 17 months to begin policy management. The model shows a minimum of 2 years is required to pass through all phases of the fork model at least once.

Future iterations of the fork model are on a 1-year cycle. Management's commitment to transformation has been demonstrated by passing through one cycle of the fork model. Hence, the handle of the fork model is utilized only on an as-needed basis. Management's education in respect to Quality Management continues indefinitely into the future. There is no fixed schedule for it. It just happens when it is deemed necessary by a manager in need of training, the manager's supervisor, the EC, or the PDC. Likewise, daily management, cross-functional management, and the initial Presidential Review portion of policy management (step 28) continue indefinitely into the future. However, steps 29 through 34 of policy management take on a yearly cycle. For example, step 29 (Policy Setting — Strategic Objectives) takes approximately 1 month, step 30 (Policy Setting — Improvement Plans) takes approximately 1 month, step 31 (Policy Deployment) takes approximately 1 month, step 32 (Policy Implementation) takes approximately 6 months, step 33 (Quality Feedback and Review) takes about 2 months, and step 34 (Presidential Review) takes about 1 month

Template

Phase 1: The Handle — Management's Commitment to Transformation

Step	Time Frame	Responsibility	Outcomes
Step 1: P creates a crisis to generate the energy for transformation.	Early-month 1	P	List of crises.
Step 2: P creates a vision to generate the energy for transformation.	Early-month 1	P	Vision statement.
Step 3: P initiates transformation using a crisis or a vision.	Mid-month 1	P	Publication of crisis and/or vision.
Step 4: P contacts an external expert in the system of profound knowledge (LC).	Mid-month 1	P	Retain LC.
Step 5: Window of opportunity for transformation opens.	Mid-month 1	P	Communication with all stakeholders about QM.
Step 6: P and LC collect data for transformation plan.	Months 1–2	P LC	Results of "barriers against" and "aids for" study.
Step 7: P and LC begin planning transformation.	Month 3	P (support and review) LC	Transformation plan.
Step 8: P forms the EC.	Month 3	P LC EC	EC is formed.
Step 9: LC trains and educates EC and future QM experts.	Months 4–6	LC EC members QM experts	Completion of training program with mastery by EC.

Template (continued)

Phase 1: (continued) The Handle — Management's Commitment to Transformation

Step	Time Frame	Responsibility	Outcomes
OPTIONAL: EC selects individuals to become QM experts by pursuing an M.S. degree in QM. These people study for 1.5 to 2 years and come on line after the first review by the P (see step 34 of the model). One QM expert per 500 employees.	Months 5–24	EC members LC QM experts University program	Completion of MS in QM by QM experts LC assists EC in selecting a university program.
Step 10: Window of opportunity for transformation begins to close without action from EC.	Month 7 and beyond	EC members LC	Communication with all stakeholders about QM process.

Phase 2: The Neck — Management's Education

Step	Time Frame	Responsibility	Outcomes
Step 11: EC forms education and self-improvement groups.	Month 8 and beyond	LC EC	1. EC answers 66 questions. 2. EC prepares Executive Summaries. 3. EC role plays . 4. EC uses new paradigms to create win–win scenarios.
Step 12: EC establishes a life-long process for education and self-improvement.	Month 11 and beyond	LC EC	LC develops a learning and self-improvement plan for each EC member.
Step 13: EC working with LC to resolve individual issues which create barriers to transformation.	Month 11 and beyond	LC EC	EC resolves concerns with QM via "inventory" tool used by LC.

Template (continued)

Phase 3: Prong 1 — Daily Management

Step	Time Frame	Responsibility	Outcomes
Step 14: EC selects initial PILs.	Month 8	LC EC PILs	Selection of initial team leaders.
Step 15: LC trains initial PILs.	Month 8	LC Initial PILs	Train initial PILs in Tools and Methods for QI and Team Methods for QI.
Step 16: Members of the EC evaluate the initial process improvement projects (daily management issues).	Month 8	LC EC Initial PILs	Initial projects selected.
Step 17: EC members, in consultation with the team leader, select the initial process improvement team members.	Month 8	LC EC members Initial PILs Initial PITMs	Teams are selected for each project.
Experts train team members.	Months 8 and beyond	Initial PILs Initial PITMs	Team members are trained.
Step 18: Initial process improvement teams conduct daily management using the QI story format.	Month 11 and beyond	Initial PILs Initial PITMs	QI story.
Step 19: Over time, other process improvement teams are formed to improve daily management.		EC members New PILs New PITMs	QI stories.
Experts train new team leaders and members together.		LC New PILS New PITMs	New team leaders and members are trained.
Step 20: LSTs coordinate daily management projects.	Month 8 and beyond	LST members PILs	QI stories.

Template (continued)

Phase 4: Prong 2 — Cross-Functional Management

Step	Time Frame	Personnel	Outcome
Step 21: Members of the EC evaluate initial cross-functional projects.	Month 12 and beyond	EC members LC	Selection of cross-functional projects.
Step 22: Members of the EC evaluate the initial cross-functional project leaders.	Month 12 and beyond	EC members LC Initial CFPLs	Selection of cross-functional team leaders.
Step 23: Experts train initial cross-functional project leaders.	Month 13 and beyond	LC Initial CFPLs	Initial cross-function team leaders are trained in (1) QM Theory, (2) Tools and Methods of QI, and (3) Team Methods for QI.
Step 24: EC members, in consultation with the team leader, select the initial cross-functional team members.	Month 13 and beyond	EC LC Initial CFPLs Initial CFTMs	Initial cross-function team members are trained in (1) QM Theory, (2) Tools and Methods of QI, and (3) Team Methods for QI.
Experts train team members.		LC Initial CFTMs	
Step 25: Initial cross-functional teams improve cross-functional issues using the "System of Profound Knowledge."	Month 14 and beyond	Initial CFPLs Initial CFTMs	QI stories.
Step 26: Over time, other cross-functional teams may be formed to improve cross-functional issues.	Month 17 and beyond	EC LC	New cross-functional teams are formed.

Step	Time Frame	Personnel	Outcomes
Other cross-functional team leaders and members are trained by LC.		LC New CFPLs New CFTMs	Cross-functional leaders and members are trained in (1) QM Theory, (2) Tools and Methods of QI, and (3) Team Methods for QI. QI stories.
Step 27: EC coordinates cross-functional projects.	Month 14 and beyond	EC CFPLs	

Phase 5: Prong 3 — Policy Management

Step	*Time Frame*	*Personnel*	*Outcomes*
Step 28: Conduct initial Presidential Review	Month 8 and beyond	P EC LC Selected PILS and PITMs CFPLs and CFTMs	Constructive critique of selected process improvement teams by the P.
Step 29: Policy setting: EC develops initial strategic objectives.	Month 11 and beyond	EC LC	Strategic Objectives.
Step 30: Policy setting: Policy Deployment Committee develops improvement plans.	Month 13 and beyond	PDC LC	Improvement plans for all areas.
Step 31: Policy deployment: PDC communicates projects to LSTs.	Month 15 and beyond	PDC LSTs PILS and PITMs CFPLs and CFTMs LC	LSTs receive and work on QI stories.
Local teams conduct projects.			

Template (continued)

Phase 5: (continued) Prong 3 — Policy Management

Step	Time Frame	Personnel	Outcomes
Step 32: Policy implementation.	Month 15 and beyond	PDC LST PILS and PITMs CFPLs and CFTMs	Findings of QI stories are implemented.
Step 33: Quality feedback and review.	Month 19 and beyond	PDC LST LC	All QI stories are reviewed by LSTs. Selected QI stories are reviewed by PDC and EC members.
Step 34: Presidential Review.	Months 22–24	P EC LC Selected PILs and PITMs	Selected QI stories are reviewed by the P.
QM experts come on line in the QM process.	Month 25 and beyond	QM experts	QM experts facilitate system-wide promotion of QM activities.

SUMMARY

This chapter presents a template for answering questions about the time and resources required to promote the fork model. Each application of the template requires the user to modify it for his or her organization. The time and cost structure is largely a function of the effort the organization devotes to the Quality Management p.ocess versus the effort it requires the consulting organization to devote to the Quality Management process.

INDEX